Water - Efficient Landscape

GUIDELINES

Richard E. Bennett
and Michael S. Hazinski

American Water Works Association

Copyright © 1993
American Water Works Association
6666 West Quincy Ave.
Denver, CO 80235

Printed in USA

ISBN 0-89867-679-7

Printed on recycled paper

Contents

Preface, v

Acknowledgments, vii

Chapter 1 Promoting Urban Landscape Water Efficiency 1

    Statement of the Problem, 2
    Purpose of Water-Efficient Landscape Guidelines, 3
    Potential for Water Conservation, 4
    Scope and Limitations, 6
    References, 7

Chapter 2 Developing Local Standards 8

    Approaches to Landscape Standards, 8
    Creating a Viable Approach, 15

Chapter 3 Applicability of Standards 18

    Scope of Implementation, 18
    New Installations, 19
    Exclusions From Applicability, 20
    Existing and Rehabilitated Landscapes, 21
    Reclaimed Water, 21
    References, 22

Chapter 4 Water Budget Approach 39

    Definitions, 39
    Historic Basis Versus Actual Landscape Water Requirements, 41
    Water Budget Factors, 41
    Water Budget Formulas, 49
    References, 52

Chapter 5 Checklist Approach 53

    Landscape Design Standards, 53
    Irrigation Standards, 58
    Maintenance Standards, 67
    Scheduling Standards, 70
    References, 71

Chapter 6 Implementing Landscape Standards 73

    Implementing the Water Budget Approach, 74
    Implementing the Checklist Approach, 76
    Educational and Informational Services, 80

Appendix A Selected Evapotranspiration and Rainfall Data 82

Appendix B Water Measurements, Factors, and Formulas for Calculating Water Requirements, Water Budgets, and Irrigation Schedules 95

Appendix C Examples of Landscape Standards 103
    Sample Checklist Approach, 104
    Sample Combined Checklist and Water Budget Approach, 124
    Sample Water Budget Approach, 152

Appendix D State Resources for Landscape Standards Information 159

Additional Sources of Information, 161

Glossary, 163

Index, 170

Preface

Landscape water efficiency is a relatively new area of concern for water purveyors and the landscape industry. In the past 15 years, water efficiency has evolved from an issue receiving marginal attention to a dominant theme for the various industries and professions that rely on water for landscape irrigation.

Increased demand and environmental considerations have permanently changed the water-resource picture. Conservation is now considered a source of water supply and is no longer just a response to short-term water shortages. The water industry, charged with providing low-cost water, is looking to improved landscape water efficiency as a demand-side component of long-term water-supply management. And, in some states, water purveyors are required to have water-conservation programs in place. These programs include regulatory and educational measures aimed at reducing water use for urban landscape irrigation.

This book, written from a water industry perspective, is intended to serve as a resource for individuals, in the private and public sectors, working to improve landscape water efficiency. The need for a comprehensive manual follows from the emphasis on demand-side water management and its impact on private industry. Water agencies can benefit from prior experience in developing and implementing landscape water-efficiency standards. The green industry can better respond to regulatory and economic pressures resulting from a changing water-resource picture. Hopefully, this book will engender a productive dialogue between the private and public sectors by serving as a framework for discussion and by clarifying the confusing maze of terminology related to landscape water use.

In 1990, the AWWA Xeriscape Committee established the development of landscape guidelines as a project priority. The authors wish to thank the members of the AWWA Water Conservation Committee and the Xeriscape Subcommittee for their commitment and support to this project. A special thanks to the East Bay Municipal Utility District (EBMUD) in Oakland, Calif., for the generous contribution of staff time and district resources to the development of this project, and to Bruce K. Furgeson of the University of Georgia for his thorough review of the manuscript.

Research conducted for this book included a review of landscape guidebooks and regulations and a search for literature addressing landscape water efficiency. Publications of the Cooperative Extension, University of California were an objective and definitive source of information. The experience of individuals within EBMUD and other agencies currently administering water-conservation programs was invaluable.

The fundamental principles of landscape water efficiency are well established, and research for a better understanding is ongoing. The application of those principles and the realization of actual water savings are essential to sustain the urban landscape. It is hoped this book will be useful to a wide range of people in the landscape and water industries.

ABOUT THE AUTHORS

Richard E. Bennett is the Water Conservation Administrator for the East Bay Municipal Utility in Oakland, Calif., and currently chairs the AWWA Xeriscape Subcommittee responsible for the development of this book. He holds a degree in Meteorology from San Jose State University and has 23 years' experience in the water and wastewater fields. He has administered water-conservation programs for the past 10 years and has been very active in the development and implementation of regulatory, incentive, and informational programs for urban water conservation.

Michael S. Hazinski is a Water Conservation Representative for the East Bay Municipal Utility District. He holds a degree in Industrial Arts from San Francisco State University where his studies emphasized industrial research and development. Trained in horticulture and landscape design, his experience includes work in residential landscape maintenance and in landscape design with Omi Lang Associates, Landscape Architects in San Francisco, Calif. His water conservation experience includes implementation of landscape and industrial programs and development of informational materials.

Acknowledgments

Thanks to the AWWA Water Conservation Committee:

Jeanne F. McKeever, Committee Chair, City of Portland Water Works, Portland, Ore.
Jeffery P. Featherstone, Committee Vice Chair, Delaware River Basin Commission, West Trenton, N.J.
Bruce Adams, South Florida Water Management District, West Palm Beach, Fla.
Damann L. Anderson, Ayres Association, Tampa, Fla.
Lynn E. Anderson, Santa Barbara County Water Agency, Santa Barbara, Calif.
Thomas M. Babcock, Phoenix Water Conservation, Phoenix, Ariz.
Brian Barrett, Waukesha Water Utility, Waukesha, Wis.
Duane D. Baumann, Planning and Management Consultants, Ltd., Carbondale, Ill.
Richard E. Bennett, East Bay Municipal Utility District, Oakland, Calif.
Juan Borges, Dade County Department of Environmental Resources, Miami, Fla.
Ron W. Brown, Metropolitan Water District of Southern California, Los Angeles, Calif.
Brenda J. Chapman, Ken Chapman and Associates, Scottsdale, Ariz.
Wendy L. Corpening, W.L. Corpening and Associates, Avery, Calif.
Duane S. Cutler, Waterguide, Inc., Montague, N.J.
William Cutler, Niagara Conservation Corporation, Flanders, N.J.
Mary Ann Dickinson, Regional Water Authority, New Haven, Conn.
Allan J. Dietemann, Seattle Water Department Water Conservation Office, Seattle, Wash.
Cynthia Dietz, Portland Water Bureau, Portland, Ore.
Edward Duerr, Brown and Caldwell, Denver, Colo.
Cynthia Dyballa, USEPA, Washington, D.C.
Benedykt A. Dziegielewski, Southern Illinois University, Carbondale, Ill.
Gary Fiske, Barakat and Chamberlin, Inc., Oakland, Calif.
John Flowers, USEPA, Washington, D.C.
John Fregonese, Department of Community Development, Ashland, Ore.
Lonnie Frost, Town of Gilbert, Gilbert, Ariz.
Larry S. Galowin, National Institute of Standards and Technology, NVLAP, Gaithersburg, Md.
Patrick J. Gleason, JM Montgomery Consulting Engineers, Lake Worth, Fla.
Mary M. Goldberg, Denver Water Department, Denver, Colo.
Alice J. Darilek Grisham, New Mexico State Engineer Office, Santa Fe, N.M.

Tony T. Gregg, City of Austin, Austin, Texas
Richard Harasick, Department of Water and Power, Los Angeles, Calif.
Rick L. Harmon, American Water Works Association, Denver, Colo.
Elizabeth Seymore Inman, Denver Water Department, Denver, Colo.
Gary F. Kah, Agtech Associates, Inc., Redwood City, Calif.
Michael King, Synergic Resources Corporation, Oakland, Calif.
Douglas J. Kobrick, Black and Veatch Engineers, Phoenix, Ariz.
Thomas Konen, Stevens Institute of Technology, Davison Research Laboratory, Hoboken, N.J.
James D. Kuykendall, Mammoth County Water District, Mammoth Lake, Calif.
Barbara S. Lahage, Massachusetts Water Resources Authority, Boston, Mass.
Mariann Long, City of Pasadena, Pasadena, Calif.
William E. Luta, Pennsylvania American Water Company, Camp Hill, Pa.
William O. Maddaus, JM Montgomery Consulting Engineers, Walnut Creek, Calif.
George Martin, Los Angeles Department of Water and Power, Los Angeles, Calif.
Fox McCarthy, Cobb County/Marietta Water Authority, Marietta, Ga.
Sue F. McCormick, Lansing Board of Water and Light, Lansing, Mich.
L.D. McMullen, Des Moines Water Works, Des Moines, Iowa
Colin Milne, Resources Conservation, Inc., Greenwich, Conn.
Jonas Minton, California Department of Water Resources, Sacramento, Calif.
Ann Marie Mitroff, City of Santa Cruz Water Department, Santa Cruz, Calif.
Faith Mueller, City of Scottsdale, Scottsdale, Ariz.
John O. Nelson, North Marin Water District, Novato, Calif.
Wendy Nero, Tampa Water Department, Tampa, Fla.
Kent Newland, City of Phoenix, Phoenix, Ariz.
Steven L. Olson, Arizona Department of Water Resources, Phoenix, Ariz.
Richard Owen, Southwest Florida Water Management District, Brooksville, Fla.
Thomas E. Pape, Volt, Orange, Calif.
Fraser Parsons, City of Edmonton, Edmonton, Alberta
Michael L. Personett, Texas Water Commission, Austin, Texas
Charles W. Pike, California Department of Water Resources, Sacramento, Calif.
Jane Ploeser, City of Phoenix Water Conservation, Phoenix, Ariz.
Marsha Prillwitz, California Department of Water Resources, Sacramento, Calif.
Philip Regli, City of Scottsdale Water Resources Department, Scottsdale, Ariz.
Alan Roberson, American Water Works Association, Washington, D.C.
Ann Roberts, The Roberts Company, Inc., Carmel-by-the-Sea, Calif.
James E. Robinson, University of Waterloo, Waterloo, Ontario

Ronald B. Rodenhaver, El Paso County Water Authority, El Paso, Texas
Dan Rodrigo, Metropolitan Water District of Southern California, Los Angeles, Calif.
David Schultz, City of Glendale, Glendale, Ariz.
Llana H. Sherman, Resource Management Agency, Ventura, Calif.
Douglas M. Short, Belmont County Water District, Belmont, Calif.
Claudia Smith, Massachusetts Water Resources Authority, Boston, Mass.
Robin Stinnett, Arizona Municipal Water Users Association, Phoenix, Ariz.
Dietrich J. Stroeh, CSW/Stuber-Stroeh Engineering Group, Inc., Novato, Calif.
Donald Tate, Environment Canada, Inland Water Directorate, Ottawa, Ontario
Jeff Taylor, Brown and Root, Inc., Houston, Texas
Harold W. Thompson, USEPA Region VIII, Denver, Colo.
David D. Todd, City of Fresno Water Division, Fresno, Calif.
Amy Vickers, Amy Vickers and Associates, Boston, Mass.
Charles W. Walker, Rosenberg and Associates, Los Angeles, Calif.
Kathy Whalen, Abracadabra Education Entertainment, Tucson, Ariz.
Trish Williamson, Tucson Water, Tucson, Ariz.

Thanks to the AWWA Xeriscape Subcommittee:

Richard E. Bennett, Committee Chair, East Bay Municipal Utility District, Oakland, Calif.
Bruce Adams, South Florida Management District, West Palm Beach, Fla.
Gary F. Kah, Agtech Associates, Inc., Redwood City, Calif.
Fox McCarthy, Cobb County/Marietta Water Authority, Marietta, Ga.
Faith Mueller, City of Scottsdale, Scottsdale, Ariz.
Wendy Nero, Tampa Water Department, Tampa, Fla.
Kent Newland, City of Phoenix, Phoenix, Ariz.
Richard Owen, Southwest Florida Water Management District, Brooksville, Fla.
Robin Stinnett, Arizona Municipal Water Users Association, Phoenix, Ariz.

A special thanks to the East Bay Municipal Utility District in Oakland, Calif., for the generous contribution of staff time and district resources to the development of this project.

We are also grateful to the following individuals for their contributions to the development of this project.

For review and comment on the manuscript:

Erika Aschmann, East Bay Municipal Utility District, Oakland, Calif.
Jacques Debrà, City of Davis, Davis, Calif.
Cynthia Dietz, Portland Water Bureau, Portland, Ore.
Bruce K. Furgeson, University of Georgia, Athens, Ga.
Rick Hornbeck, East Bay Municipal Utility District, Oakland, Calif.

Fox McCarthy, Cobb County/Marietta Water Authority, Marietta, Ga.
John O. Nelson, North Marin Water District, Novato, Calif.
Andrea Pook, East Bay Municipal Utility District, Oakland, Calif.
Marsha Prillwitz, State of California Department of Public Works, Sacramento, Calif.
John Swindell, East Bay Municipal Utility District, Oakland, Calif.
Steven Whitehill, Water Management Resource, Cupertino, Calif.
Valerie Whitehill, Water Management Resource, Cupertino, Calif.

For word processing:

Nancy Taconni, East Bay Municipal Utility District, Oakland, Calif.

For graphic illustration:

Len Deneweth, East Bay Municipal Utility District, Oakland, Calif.
Eli Frank, East Bay Municipal Utility District, Oakland, Calif.

For assistance with the metric conversions in appendix B:

Fraser Parsons, City of Edmonton, Edmonton, Alberta

We thank the East Bay Municipal Utility District's (EBMUD) Landscape Advisory Committee, a group comprised of San Francisco Bay Area landscape professionals, for review of the manuscript.

EBMUD Landscape Advisory Committee

Buzz Bertolero, Navelet's Nursery, Walnut Creek, Calif.
Tom Courtwright, Orchard Nursery, Lafayette, Calif.
Bob Cox, Canyon Lakes Country Club, Danville, Calif.
Marty Dickson, Russell D. Mitchell & Associates, Inc., Walnut Creek, Calif.
Roger Fiske, Fiske Landscaping, Danville, Calif.
Pete Gumas, Rainbird, Pleasanton, Calif.
Lisa Hagopian, Rossmoor Homeowners Association, Walnut Creek, Calif.
Fred Hankler, Delta Bluegrass, Stockton, Calif.
Don Laughland, Professional Landscape Management, Danville, Calif.
Karen Mahshi, Mahshi Association, Oakland, Calif.
Paul McClure, Chevron, San Ramon, Calif.
Tom Raeth, Lafayette Tree and Landscape, Lafayette, Calif.
Phil Reiker, Greenfield Turf, Capitola, Calif.
Jeff Reuter, Jeff Reuter Irrigation, Walnut Creek, Calif.
Jim Salvador, Heather Farms Landscape, Pleasant Hill, Calif.
Chet Sarsfield, Irrigation Tech Services, Brentwood, Calif.
Mark Scharmer, Irrigation, Walnut Creek, Calif.

1

Promoting Urban Landscape Water Efficiency

As increased demands for limited water resources challenge supplies, improved urban water efficiency is essential for indoor and outdoor use. Urban landscape irrigation, often considered a discretionary use, is targeted by water purveyors for reduction because of its highly visible potential for water savings.

Historically, landscape design and management practices have not addressed water efficiency. Consequently, inefficient designs and management have resulted in excessive water use and high long-term maintenance costs. Increased efficiency is a positive alternative to the total elimination of irrigated landscaping because it reduces demand, yet preserves the inherent value of the irrigated landscape.

Irrigation of landscapes is not entirely discretionary since it supports important functional, recreational, aesthetic, and economic values. Functional values include fire protection, erosion control, soil stabilization, temperature modification, energy conservation through the creation of shade and windbreaks, and reduction of storm drainage flows. Recreational amenities such as parks, golf courses, and play fields are highly valued in most communities.

Aesthetic values coincide with functional and recreational values. For example, reducing glare and heat gain make a harsh urban environment more livable. And, most communities place a high priority on tree planting programs and other community-based landscape restoration programs.

As for economic values, increased real estate values are associated with an attractive landscape. Landscape and recreation industries and their associated economies depend on water for landscape irrigation.

STATEMENT OF THE PROBLEM

The present and growing need for water-efficient landscape guidelines results from existing inefficient uses of water and increased demands on limited resources. Water-resource limitations vary from region to region based on climate, demand, and historic water-resource development.

Existing Inefficiencies

Historically, landscape designers and government planning agencies have not addressed landscape water efficiency. A precedent of design for aesthetics rather than function and a historic absence of efficiency standards contribute to existing inefficiencies in landscape water use. But water efficiency in current landscape practice can be improved through developing standards and educational programs for landscape industry professionals and consumers.

Existing inefficiencies in landscape water use can readily be observed. For example, poor design of irrigation systems often results in overspray and runoff onto streets and pavement. Overspray usually occurs in the spray irrigation of narrow median or parkway strips. Runoff is usually a problem when turfgrass requiring spray irrigation is planted on slopes and berms.

A less obvious type of waste occurs when water is unevenly applied. As a result, in order to ensure that all parts of a landscape receive enough water, some parts of a landscape receive too much water.

Poor management often results in flooding due to undetected leaks and in soggy landscapes due to incorrect scheduling. Over-irrigation also can result in the loss of landscapes where plants cannot tolerate soggy soil conditions. Such waste is unnecessary and preventable. Landscape standards would help minimize such design and management inefficiencies.

The historic precedent of design for aesthetics rather than function also results in higher long-term maintenance costs. European landscape design and practice, inappropriate to many North American climate and soil conditions (especially the arid and rocky western portions of the United States), have been developed and used on a massive scale in portions of the United States and Canada. Plants that require a large amount of water and are unadapted to arid and semiarid climates are commonly used in the Western United States. Even in areas of substantial natural precipitation, it is common to find plants used in the landscape that require supplemental irrigation.

The educational gap in the landscape industry is evidenced by limited understanding of water-efficient design and management. A gap also exists between new water-efficient products and methodology and their proper application in the field. Higher efficiency is more attainable as new products and techniques become available.

The educational gap among consumers is associated with traditional landscapes, habit, and a lack of attention to consumption practices. A demand for nonfunctional, high-water-use landscaping persists despite the availability of viable, attractive, and cost-effective alternatives. This educational gap continues to contribute to the wasting of a limited resource.

Increased Demand for Water Resources

Increased demand for water resources by a growing population and its associated activities is more and more at odds with irrigating landscapes. This increased demand, combined with natural climatic fluctuations, has contributed to periodic water shortages in many states. These shortages heighten public awareness of the need to improve efficiency and generate momentum for regulatory action.

The benefits of landscape water conservation should be measured against the cost of developing additional supplies of water. Because landscape irrigation is often a large component of average, and especially peak, demand for a water utility, landscape water-conservation programs may reduce or forestall the development of new and expensive water supplies. And, successful water conservation programs can result in lower long-term cost to customers.

PURPOSE OF WATER-EFFICIENT LANDSCAPE GUIDELINES

The primary purpose of landscape guidelines is to promote water efficiency in the urban landscape. Efficiency standards that reduce demand will contribute to short-term conservation efforts during drought emergencies and to long-term water-supply management.

Guidelines can be implemented through educational programs and through adoption of regulations by water utilities and/or government agencies. Many communities have recognized the potential for conservation in the landscape and have developed appropriate landscape regulation. State and local standards also have been adopted that regulate other water uses, including indoor plumbing. Landscape standards can function as a component in more comprehensive water-conservation programs.

Because water for landscape irrigation serves a single purpose (maintenance of a healthy and attractive landscape), landscape irrigation provides a unique opportunity to establish efficiency standards for a major category of water use. By contrast, industrial and commercial water use is varied and serves many purposes. As a result, efficiency standards are difficult to establish for these categories of use. However, the generic nature of landscape water use makes efficiency standards viable.

A comprehensive set of guidelines should be based on specific objectives that support the primary purpose of maximizing efficiency. These include

- adapting to local water-resource limitations
- reducing water requirements for new landscapes
- increasing efficiency of existing landscape design and management
- educating landscape professionals, regulators, and the general public

These objectives may be weighted differently, based on local opportunities and constraints.

Adapting to Local Water-Resource Limitations

Maximizing efficiency is best achieved on a local level where programs can be integrated into more comprehensive conservation goals and programs. Because consumption patterns vary regionally, locally developed landscape guidelines can address uses that hold the maximum potential for water conservation. Water-resource limitations will determine the need for guidelines, and local patterns of use will determine the level of implementation required.

Reducing Water Requirements for New Landscapes

A tremendous opportunity exists to improve the design and management of new landscape installations. Approval of new landscape projects and water-service connections for irrigation can be made to be contingent upon meeting specific criteria for water efficiency. Practical, functional, attractive, and sustainable landscapes can be achieved through better design and management.

Increasing Efficiency of Existing Landscape Design and Management

Much of the existing water waste is due to inefficiencies in irrigation systems and improper scheduling that is not responsive to weather or actual water requirements for plants. Standards for efficient maintenance and scheduling can significantly reduce water consumption, especially where automated systems are in place. Because existing landscapes require periodic renovation, a potential also exists to reduce demand through improved redesign.

Education

The demand for nonfunctional, high-water-use landscaping may be reduced by educating the public to positive, water-efficient alternatives. Most consumers will respond favorably to increased cost-effectiveness of maintaining water-efficient landscapes. Benefits to the environment, including the preservation of native plant communities, can influence consumer choices in the development and maintenance of landscapes, as well.

In addition, government agencies, as water users, can provide leadership in water-conserving landscape and irrigation practices. For instance, agencies can adopt water-efficient landscape standards for their own landscape irrigation activities.

Regulatory programs have educational value because they prescribe criteria for water efficiency. Water utilities, local government agencies, and private consultants can implement educational programs, too. Supplemental information and materials, tailored to local needs and constraints, will

- facilitate the acceptance of regulatory programs
- help landscape professionals identify cost-effective water-conservation opportunities and improve their water-management skills
- facilitate application of new water-conserving technology and methodology in the field

POTENTIAL FOR WATER CONSERVATION

The potential for landscape water conservation will vary significantly by region (Figure 1-1). In the arid western states, 40 to 50 percent of residential water consumption is used for landscape irrigation. In climates characterized by annual drought and a rainy season, many landscapes rely almost entirely on irrigation for most months of the year; water use can be two to four times the actual plant requirements as measured by evapotranspiration rates.

In climates of the East, Southeast, and Northwest, which have substantial rainfall, irrigation use for landscapes will be significantly less. But increased demand for landscape irrigation during episodic drought can contribute to periodic water shortages. In these climates, water-efficiency standards that require regionally adapted plants and efficient irrigation will reduce peak demand during periods of reduced supply.

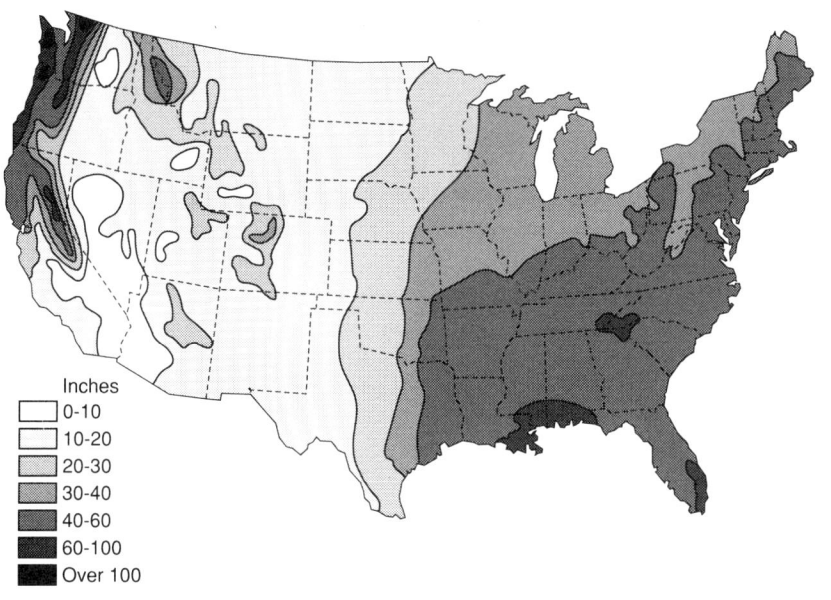

NOTE: Values for any given area are often misleading. Wide variability in rainfall occurs seasonally, annually, and, in some instances, in areas of proximity.

Source: *US Water Resources Council (1978)*.

Figure 1-1 Average annual precipitation for the United States

Conservation through improved efficiency is maximized when all areas of landscape practice are addressed. Comprehensive landscape guidelines will encompass

- local water-conservation opportunities
- appropriate design
- appropriate irrigation technology
- efficient management practices

Local Opportunities to Conserve Water

Opportunities to conserve water depend on the local climate, terrain, soils, and patterns of use. Inefficiencies predominant in a given area need to be identified. For example, sprinkler irrigation will generate wasteful runoff when applied to hilly terrain and clay soils. In communities where these conditions are common, guidelines and standards should address this inefficiency.

Types and patterns of water use will vary greatly from one area to another. For example, landscape irrigation in densely populated urban settings may be concentrated in large city-owned parks and landscapes, while residential use may predominate in suburban settings. Water-efficiency criteria that are specific to the type of user would maximize potential savings.

Appropriate Design

Landscape design can significantly increase potential water efficiency in the landscape. Inefficient or inappropriate design, including nonfunctional water-consumptive landscaping, selection of plants requiring a lot of water, and wasteful irrigation systems, contributes to a landscape's water demand. These elements can be reviewed as to whether they comply with the predetermined criteria, thereby reducing landscape water requirements.

Appropriate Irrigation Technology

Development of irrigation technology and methodology by university researchers and private industry has increased the potential for water efficiency in the landscape. Technological advances in sprinkler and drip irrigation and in irrigation control imply higher standards for efficiency. However, consumers and landscape professionals suffer from an information gap over products and their effective application. Again, guidelines and standards would facilitate the proper application of water-conserving technology.

Efficient Management Practices

Designing water-efficient landscapes does not ensure efficient application in the field. Irrigation scheduling that does not meet actual climate and plant water requirements is a major source of excessive consumption. Landscape managers tend to err on the side of over-irrigation to maintain appearance, especially of turfgrass.

In fact, over-irrigation is the most important area to address when establishing guidelines or regulation. Actual landscape water requirements can be determined in the design stage of a project. Then long-term irrigation schedules can be prepared based on average climate and plant water-requirement data.

Another source that contributes to inefficient water use is improper maintenance of both landscapes and irrigation systems. Maintenance procedures that increase potential water efficiency also can be specified in the design stage.

SCOPE AND LIMITATIONS

This book is intended as a resource for landscape industry professionals and agencies working to improve water efficiency in the landscape. Information within this book can be used as guidelines or as a tool to develop regulations.

A note of caution: It is not practical to develop a set of guidelines applicable on a national level. Local variations in climate and geography prescribe the relevance of an approach and specific criteria. Furthermore, the implementation of regulatory approaches must be suited to an agency's administrative resources. A specific set of national standards may not accommodate the administrative capacity of local jurisdictions.

While most specific criteria are not applicable nationally, some uniformity of standards within a region is desirable. Regulatory variation in local jurisdictions may cause confusion for landscape professionals who work in a range of locations. Standardized regional criteria will improve compliance with regulatory programs and facilitate the exchange of information between landscape professionals and the enforcing agency.

Scope

The following topics are presented later in this book as a comprehensive overview of issues and approaches to water-efficient urban landscapes.

Basic approaches. Basic approaches to improving water efficiency are defined and their advantages and disadvantages discussed. Some of the most common approaches include the following:

- A water budget approach defines limits on consumption.
- A checklist approach prescribes water efficiency through specific design and management criteria.
- An educational approach provides informational materials and services and can supplement regulatory programs.

This book discusses methods of combining these basic approaches as well. Sample programs and technical information are presented as appendixes.

Applicability. This book addresses the applicability of landscape guidelines and standards to various types of landscape development. Defining the range of applicability, especially of regulatory standards, is essential to their effectiveness.

Specific elements. Various elements that contribute to water efficiency in the landscape are presented in a more specific discussion of water budget and checklist approaches. These can be evaluated for their appropriateness to a given locality. Elements include

- landscape design—grading, irrigation, soils, plant selection, and grouping
- landscape management practices—irrigation scheduling and maintenance, grounds maintenance, and horticultural practice

How to implement regulatory and educational programs. This book also discusses implementation mechanisms for regulatory and educational programs. Successful programs depend on appropriate standards and effective enforcement. Design criteria alone will not guarantee water efficiency in the landscape because efficiency also depends on correct, ongoing management practices.

Limitations

A specific set of national guidelines are not delineated since appropriate methods of improving efficiency depend on local conditions. The advantages and disadvantages of each approach are presented to facilitate the adoption of guidelines that meet local needs.

Since landscape water efficiency is a new and evolving area of study, more research and field testing is necessary to quantify and document efficiency standards. Quantitative standards are intended to be flexible and are subject to further scientific study.

Implementation of regulatory approaches can be facilitated through water-conserving rate structures. The design of rate structures, however, is beyond the scope of this publication.

The cost of water for irrigation will vary greatly within local regions. The monetary benefit of investing in water-efficient design and management is an important area of consideration and is related to incentives for efficiency created by local water rates. Cost–benefit analysis of landscape water-conservation measures is site-specific and is also beyond the scope of this publication.

REFERENCE

US Water Resources Council. 1978. *The Nation's Water Resources, 1975–2000.* Second National Water Assessment. Vol. 1: Summary. Superintendents of Documents, Washington, D.C.

2

Developing Local Standards

APPROACHES TO LANDSCAPE STANDARDS

Setting up water-efficiency guidelines can be approached from three different angles (Table 2-1). Two distinct approaches can be discerned from existing ordinance and regulation. These are the water budget approach and the checklist approach. The former describes limits on water consumption and the latter prescribes design and management criteria. A combined water budget and checklist may best meet local needs. A third approach, the educational approach, may serve as an alternative to regulatory approaches or can be used in conjunction with another approach to facilitate the implementation of water-efficiency standards.

Water Budget Approach

A water budget approach establishes limits on allowable consumption in an irrigated landscape. Consumption limits are defined by a formula representing efficiency standards applied to a site. (The formula does not specify design and management criteria as is done in the checklist approach.) A water budget is a site-specific quantity of water needed to maintain a healthy landscape. Annual water budgets should be distributed throughout the year in monthly or bimonthly allocations that reflect seasonal variations in water requirements.

An effective method requires the user to present a documented water budget. Implementation requires an enforcing agency, landscape designers, or landscape managers to set appropriate water allocations and monitor consumption. Mandatory standards require enforcement of consequences for exceeding the water budget. Voluntary approaches rely on economic, aesthetic, and environmental benefits as incentives.

Landscape water requirements can be estimated by using data on water requirements for local climate and plants. A site-specific water budget is determined using a formula that includes evapotranspiration, irrigated area, and irrigation efficiency factors and other variables, which are discussed in chapter 4. The means by which the user achieves efficient design and management are not prescribed in this approach.

Advantages. The water budget approach offers several advantages, which are discussed in the following paragraphs.

Price incentives. Price incentives as an enforcement mechanism can be an effective means of reducing water consumption. Increased charges for excess use make an investment in the upgrade of irrigation equipment and better landscape management more cost-effective. (Large landscapes with high water costs, such as golf courses, already are more likely to be well managed because of the potential for reduced operating costs.) Water purveyors also have used price incentives successfully to achieve general water-conservation goals in drought emergencies.

Design flexibility. Design flexibility is possible because a water budget does not advocate specific design solutions. The designer and consumer are allowed flexibility while being required to achieve a general standard of efficiency. For instance, restrictions on elements that use a large amount of water, such as water features and turfgrass, are avoided while requiring overall efficiency.

Water-efficient landscape management. Water budgets require monitoring of the degree and effectiveness of water applications. Too often, billing and consumption information are not made available to landscape maintenance personnel. These professionals also are not held accountable for efficient water management. Since a water budget represents an efficiency standard, excessive use of water can be identified. Seasonal or monthly water allocations promote scheduling based on actual requirements.

Existing enforcement mechanisms. An enforcement mechanism is usually in place. Where irrigation use is metered for billing purposes, a structure for monitoring consumption and enforcing limits is already in place. This helps minimize the plan submittal and review process because specific design criteria are not established as in the checklist approach.

Water budget review consists of calculating and documenting water allowances based on the square footage of an irrigated area. Once water allowances are established, compliance can be monitored through the billing system. This reduces the need for on-site water auditing and review of scheduling and maintenance. Water-audit services, however, should be provided to assist customers in achieving water-conservation goals.

Disadvantages. The water budget approach does have several disadvantages.

Ongoing monitoring. A water budget approach with no specific standards for design and management is enforceable only through an ongoing water-monitoring program administered by an enforcing agency. A water budget approach may be impractical for planning agencies to implement if the necessary consumption data is unavailable or a water purveyor is unable to participate. Planning and building agencies responsible for enforcing landscape water-efficiency standards may need to use a checklist approach.

Dedicated metering. Implementation of water budgets requires irrigation use to be separately metered. Unfortunately, irrigation use is often combined with other nonlandscape uses on the same meter. Separate irrigation meters or submetering of the attendant irrigation system is necessary to monitor irrigation use. Dedicated metering is an additional cost and may include connection fees and system capacity charges. In addition, separate meter reading procedures and the handling of

Table 2-1 Approaches to establishing landscape guidelines

Approach	Definition	Enforcement Mechanism	Recommended Implementing Agency	Applicability	Advantages	Disadvantages
Water Budget	Establishes limits on consumption for irrigated landscape installations	Water service contingencies Rate structures Penalties	Water utilities Planning agencies with cooperation from water purveyor	Metered irrigation service Submetered irrigation	Price incentives are effective. Allows design flexibility rather than mandating specific design solutions. Promotes water-efficient landscape management. Enforcement mechanism is usually already in place where irrigation use is metered.	Requires ongoing monitoring and cooperation of water purveyor. Requires dedicated metering of irrigation use. May create inequities among different types of consumption. May have a negative public image because of penalties and consumption limits.
Checklist	Prescribes design and management criteria	Design review Site inspection Water auditing Penalties	Local government planning agencies Sign-off by licensed and certified landscape professionals	New and rehabilitated landscape: • Residential • Commercial • Industrial • Public	Promotes water-efficient landscape design. Checklists have educational value. Does not require dedicated metering and ongoing monitoring of consumption. May be easier for public agencies to implement; does not require data on irrigation consumption.	Reduces design flexibility and may inadvertently restrict viable design solutions. Difficult to standardize requirements over a wide geographic range. May unduly encumber small projects. Difficult to address landscape management through checklists.

Table continued next page.

Table 2-1 Approaches to establishing landscape guidelines (continued)

Approach	Definition	Enforcement Mechanism	Recommended Implementing Agency	Applicability	Advantages	Disadvantages
Education	Dissemination of information	Voluntary compliance	Local government entities	Landscape professionals	Increases knowledge of water-efficient landscape practice.	Weak incentives mean reduced effectiveness.
		Licensing and certification programs	Water utilities	Homeowners	Promotes the intrinsic benefits of water efficiency.	Does not fulfill state mandates.
			University extension services		Requires minimal government intervention and administrative costs.	
			Professional associations			
			Private consultants		Promotes public acceptance of regulatory programs.	

DEVELOPING LOCAL STANDARDS 11

water-flow charges for submeters may complicate administrative procedures for recording and billing consumption.

Inequities. Inequities between different types of consumption may occur. Separate rate structures and penalties applied to irrigation use may single out this type of consumption, which is not entirely discretionary. Unless landscape water budgets are part of a more comprehensive water-conservation program that includes industrial, commercial, and residential use, inequities could be created.

Possible negative image. This approach may carry a negative image. A water budget approach may subject customers to an ongoing rate structure or surcharge that penalizes excessive consumption. Water service also may be contingent upon the customer staying within the limits of water allocations. Long-term consumption limits and the increased rates often associated with rationing programs may be negatively perceived. Without awareness of the need for long-term water conservation, a water budget program may meet resistance from the landscape industry and consumers.

Checklist Approach

The checklist approach prescribes specific design criteria for grading, soil amendment, planting, and irrigation systems. It also lays out programs for irrigation scheduling and landscape maintenance. Estimates of landscape water requirements or irrigation schedules may be used to quantify design efficiency.

Using the checklist approach requires design review and approval in the preconstruction stage of new projects and may be linked to the permitting process. An inspection after construction is completed if necessary to verify compliance. So, certified professionals may be required to sign off on projects or the enforcing agency may assume responsibility for plan review and site inspection. A water audit program is necessary to monitor ongoing compliance.

Advantages. The checklist approach offers several advantages, which are discussed in the following paragraphs.

Efficient design. A checklist promotes water-efficient landscape design. By complying with specific design criteria, water efficiency can be increased. Reduced consumption becomes a function of water-efficient design and proper landscape management.

Educational value. A checklist educates and directs designers and landscape managers towards the goal of water efficiency. Checklists are informative because they offer a step-by-step method for achieving water efficiency. Checklists can function as a resource for landscape designers and managers, especially if supplemented by informational materials and services.

No dedicated metering or monitoring. Checklists do not require dedicated irrigation metering and monitoring. The checklist approach requires efficiency without monitoring water consumption. However, once specific requirements have been met in the design and construction phases of a project, periodic auditing of irrigation efficiency and scheduling may be necessary.

Easy implementation. Checklists may be easier for public agencies to implement. The checklist approach does not require consumption data for implementation and may be the only practical approach available to a planning agency. Water budgets can be incorporated into a checklist approach as a design and management tool rather than as a method of fixing water-use limits. For example, a water budget can be used as one item on a checklist. Water requirements resulting from the design can be quantified, then compared to the water budget.

Disadvantages. There are some disadvantages to using checklists.

Reduced design flexibility. Design standards may inadvertently restrict viable design solutions. Advances in irrigation technology and in the hybridization of water-conserving plants can make checklists obsolete and actually impede the application of new materials and techniques in the landscape. For example, a strict requirement for rain shutoff devices on irrigation controllers could restrict the more advanced application of central irrigation control directly linked to weather stations. Also, the development of more grasses tolerant to drought and subsurface drip irrigation may make restrictions on turfgrass areas inappropriate.

Creative solutions by adept designers, able to combine aesthetics and function while minimizing water requirements, may be encumbered by specific regulatory requirements. So, checklists need to accommodate varied design solutions and be revised periodically to reflect advances in technology.

Hard to standardize. Standardizing requirements can be difficult over a wide geographic range. Standards must meet local needs. Statewide standards, much less national standards, may be impractical, depending on geographic and climatic variations. For example, variations in soil types, elevation and grades, and microclimates can complicate the application of the same standards to different conditions. Unusual land use or natural conditions can contraindicate low water use. For example, urbanization in areas prone to wildfires will benefit from well-watered lawns, rather than from trees and shrubs that require less water.

Multiple and varied standards within a region will complicate understanding and compliance by landscape professionals. Standardization of requirements, wherever possible, will facilitate compliance by the landscape industry. Local jurisdictions may be the only practical level at which checklists can be implemented. However, various local standards and enforcement could create a confusing array of regulation for design professionals who work on a national, statewide, or regional range.

Encumbers small projects. A submittal and review process in the planning and construction phases of a project would require additional professional design and consultation services. This would not be a problem for larger projects, which usually encompass these services. Additional costs could be absorbed more easily. However, smaller projects with less potential for water savings could be burdened by the additional requirements. This disadvantage can be reduced by excluding projects based on a minimum landscape area or construction costs.

Landscape management difficult. Thorough implementation requires design review, site inspection, and ongoing water auditing. Compliance with a design checklist does not verify correct installation and proper landscape management. On-site inspection and a follow-up program of water audits would be necessary.

Certified landscape professionals could verify proper installation and ongoing compliance with management criteria. This use of professional services would reduce some of the review and inspection by the enforcing agency. Many local governments do not have the trained staff or resources to fully implement a checklist approach.

Educational Approach

An educational approach to landscape standards involves the dissemination of information to landscape professionals and the general public. The need for such an approach is created by a historic neglect of water efficiency in landscapes and by developments in research and technology that make greater water efficiency possible. An educational program alone is not a comprehensive solution, but it is a necessary and significant component to improving landscape water efficiencies.

The educational approach relies on voluntary participation or can be part of professional licensing and certification programs. Government agencies, water utilities, educational institutions, professional associations, and private consultants can contribute to an educational effort.

Water-efficient landscapes benefit both consumers and suppliers of water. The economic and environmental benefits induce improved water efficiency. Offering educational services generates greater acceptance by the landscape industry and improves compliance with mandatory standards at all levels.

Effective educational programs currently operate in many states. Many water agencies offer free water audits of landscapes and water-audit training. In addition, water conservation is a primary topic at landscape industry conferences and trade shows. Information for professionals in landscape design and maintenance should supplement mandatory requirements and be available in foreign languages in multicultural regions.

Advantages. An educational approach offers several advantages.

Increases knowledge. An educational approach can increase knowledge of water-efficient landscape practice. Understanding viable design alternatives and water-efficient landscape management techniques is essential to realizing water savings. Through increased knowledge of water efficiency, landscape professionals can more effectively adapt to a changing water-resource picture. Informed homeowners and landscape managers will demand and expect improved efficiency.

Promotes intrinsic benefits. The intrinsic benefits of water efficiency can be promoted. When these benefits are known, they serve as an incentive to conserve. Landscape professionals who understand water efficiency will be more competitive and adaptable as consumers become aware of the monetary and environmental savings available through efficient landscape practices.

Reduces costs. Government intervention and administrative costs can be reduced. The regulatory process can encumber smaller landscape projects and also may place additional administrative burden on government agencies with limited resources. An educational effort may achieve some water conservation where mandatory standards are not viable.

Promotes acceptance of regulatory programs. In communities with regulatory programs, informational services will facilitate acceptance of and compliance with standards. Some confusion over new regulations is unavoidable but can be reduced by disseminating supplemental information. Presentations, classes, and informational materials are an opportunity to build cooperation.

Disadvantages. An educational approach does have two disadvantages.

Minimal effectiveness. Because it relies on voluntary efforts by the landscape industry and consumers, an educational approach will have minimal effectiveness in achieving water savings. Consumers and landscape professionals will not see immediate payoff since water costs are low and entail a long payback period to recoup investments in irrigation system upgrades or additional labor.

State mandates not met. Where states require local governments to implement water-efficiency ordinances, an educational approach will not meet the state requirements.

CREATING A VIABLE APPROACH

The water budget approach is more viable for water purveyors and agencies that have the capacity to monitor consumption. The checklist approach is more viable for planning agencies that can integrate a regulatory program into existing administrative procedures.

A combination of water budget and checklist and educational approaches may minimize disadvantages and maximize advantages inherent in each approach. The checklist approach may include water budget calculations that can replace some specific criteria and review procedures. Conversely, a water budget approach may be supplemented by a checklist to facilitate compliance. Informational services and materials also are necessary components of any program.

Criteria for Effective Standards

When developing landscape standards, it is important to consider the following issues (Table 2-2).

Standards should identify and target local conservation opportunities. The local patterns of water use will indicate where most savings can be generated. Although a program that targets fewer large irrigators will be easier to administer, many residential communities will need to target many smaller users who collectively are the primary source of landscape water use. Applicability and the type and size of landscapes to which a set of standards applies should be clearly defined in the standards.

In addition to an analysis of the users, inefficiencies in landscape design and management that are locally significant should be identified. Remedies then can be promoted through guidelines and standards. For example, a community can target inefficiencies such as inappropriate plant selection, irrigation of native plant species and plant communities, waste due to difficult terrain and soil conditions, lack of automated irrigation, or mismanagement of automated systems.

Table 2-2 Criteria and approaches for effective standards

Criteria	Approach		
	Water Budget	Checklist	Education
Identifies conservation opportunities	Difficult to include nonmetered consumption in range of applicability.	Can target opportunities through specific criteria for efficiency.	Improves users' ability to identify opportunities.
Enforceability	Through metered consumption and rate structures.	Through design review and onsite inspection.	Voluntary.
Design flexibility	No specific design criteria; greater flexibility.	Imposes some specific criteria; less flexibility.	Improves design solutions.
Addresses management	Inherent in monitoring consumption.	Requires water-audit or site-survey program.	Can improve scheduling and maintenance.
Preserves aesthetic and landscape value	Water allowance can allow for adequate maintenance.	Exceptions to some criteria may be required.	Improves awareness of attractive water-efficient alternatives.
Educational component	Requires supplemental informational services.	Some inherent educational value.	Functions as component in other approaches.

Landscape standards must be enforceable by the local agency. The first step is to assess available administrative resources so that implementation can be tailored to the enforcing agency. If landscape and irrigation professionals are allowed a voice in developing these standards, they will more likely accept and comply with the standards and ensure their pertinence for the community.

At the outset, all standards, submittal requirements, and consequences of noncompliance must be made clear to those potentially affected by them. Applications for new water-service connections and construction permits are an effective tool to help implementation. Detailed submittal forms will facilitate the exchange of information between the enforcing agency and landscape professionals.

If professional landscape design services are not part of a project, obtaining adequate planning and site information may be a problem. This is an important consideration when extending regulations to include private residential landscape development.

Standards should not encumber small projects. The potential for water conservation should be commensurate with the demands regulation places on landscape development. An extensive review process for small-scale landscape development may unduly encumber small projects while actually contributing little to water conservation goals.

Small landscape installations often are laid out in the field based on conceptual or schematic plans. A requirement to submit construction documents would introduce an unnecessary level of technical design service and cost. Small projects might best be excluded from complying with standards, based on cost or square footage.

Standards should limit the imposition of specific design solutions. Mandating specific design solutions, as in requirements for types and percentages of allowable plants and specific irrigation components, can unnecessarily restrict design options. In fact, well-intended regulations may inadvertently preclude viable solutions in unpredictable ways. For example, the hybridization of plant species, especially turfgrass, could produce more drought-tolerant varieties. Absolute limits on use of turfgrass may result in only minor water savings.

Also, performance specifications for both spray and drip irrigation are improving. Spray-irrigation heads are being developed to increase their efficiency, while new subsurface drip irrigation has become more feasible for turfgrass and ground covers.

Standards should address efficient landscape water management. A comprehensive approach should address ongoing water management in addition to landscape and irrigation design. Experience from water audits indicates the majority of waste often is due to inefficient management. Criteria for efficient design and installation are an important step but will not ensure proper management. Good management will include correct scheduling and maintenance.

Standards should preserve aesthetic and economic value inherent in the designed landscape. Actual water savings may not justify an overriding loss of function and economic value in recreational facilities and public amenities. Significant water savings can be achieved just by increasing overall efficiency. Exceptions in the requirements may be necessary for parks, cemeteries, and other functional and traditional landscapes to preserve their value to the community.

Standards should include an educational component. Informational materials and services serve an important support function in enforcing landscape standards. Benefits of water efficiency (beyond water savings), such as the preservation of native plant and animal communities and reduced pest and weed control problems in the constructed landscape, should be promoted to the public and landscape professionals. Preserving our water resources and the environment for future generations can be emphasized. Exposure to the benefits of water conservation in the landscape will create incentive to reduce consumption.

Educational efforts are especially important in the areas of irrigation design, scheduling, and maintenance. Landscape water-auditing programs help improve maintenance and scheduling. Water-auditor certification programs have been developed to train professionals to evaluate irrigation efficiency and make scheduling recommendations. Educational programs also should be extended to field personnel in charge of irrigation scheduling.

3

Applicability of Standards

Guidelines and standards can be applied to a variety of landscapes, from small residential to large commercial landscapes. Regulatory programs should define the type of sites to which standards apply.

SCOPE OF IMPLEMENTATION

Expanding the range of applicability for standards will increase the administrative structure necessary to implement them. For example, smaller, private residential customers will impose a greater administrative burden with proportionately less water savings. And, water savings can be maximized by including large irrigators in the scope of regulation.

The range of applicability should be determined by the type of standards to be enforced as well as the administrative resources at hand. A checklist approach is easier to implement for new landscape construction than for existing landscapes because efficient design can be required as part of the building permit process. Water budgets and irrigation-efficiency standards can be applied to existing landscapes, however, without requiring fundamental changes in design.

Efficiency standards for new irrigation-service connections can minimize the impact of development on water-system capacity. This has implications for long-term water-supply management. Enforceable landscape standards combined with an effective implementation program can be a cost-effective, demand-side component in water-resource planning.

NEW INSTALLATIONS

Enforcement of building codes by local government agencies involves applying for a construction permit and an inspection. Codes affecting water efficiency for landscape installations may already be established in many communities without specific landscape water-efficiency ordinances.

Water purveyors can identify opportunities to conserve water through water-service applications for landscape irrigation. Application and inspection procedures can incorporate specific water-efficiency requirements or establish consumption limits for a project. Categories of new construction that are likely to fall under the purview of landscape standards are discussed below.

Single and Multifamily Residential

Irrigation for residential landscapes may comprise a significant portion of total consumption within a local jurisdiction. However, residential landscape irrigation, especially for homeowners who do their own landscaping, is the most difficult type of irrigation to monitor.

Large numbers of residential users can make implementation of regulations impractical because irrigation use usually is not metered separately. And a submittal and review process of landscape plans for large numbers of small users may be impractical. Although agencies need to inform homeowners of the regulatory requirements, most individual customers lack the technical knowledge needed to achieve minimum levels of efficiency in their landscapes.

It may be necessary to exclude homeowners' landscapes from requirements for plan review and site inspection. General restrictions against flagrant waste such as gutter flooding, overspraying, and hosing of sidewalks may deter inefficiency. Striking a balance between range of applicability and water conservation is essential.

For such landscapes, educational programs may be a good alternative to regulatory programs. Such programs would provide a means by which private, residential users could voluntarily reduce consumption. Education without incentives, however, would limit compliance.

Developer-provided landscaping for individual homes and the common areas of housing developments is easier to regulate. For example, single-family and multiunit developments often have common landscape areas requiring irrigation. Because these common areas are usually professionally designed and managed, they are easier to review and inspect for compliance to efficiency standards. Separate irrigation metering can be required to facilitate monitoring and enforcement of consumption limits. Due to their large size, common areas also have proportionately greater potential for water savings than private, residential landscapes.

Developer-provided landscaping for individual residences entails preparation of complete construction documents. Streamlining the administrative process is possible by reviewing just a few design models for compliance to efficiency standards.

Model Homes or Temporary Development

Model homes provide an opportunity to demonstrate the benefits and aesthetic appeal of an efficiently designed landscape. High-quality professional design and installation can serve as an example to potential homeowners.

Developers may resist using water-efficient landscapes as prototypes since the landscapes may not have a traditional appearance. However, since model home landscapes typically are over-irrigated to maintain an extremely lush appearance and are highly visible, they provide an excellent opportunity to promote attractive, water-efficient alternatives.

Commercial Development

Because commercial development may include extensive irrigation, it should be included in the range of applicability. However, some criteria should be established that exclude commercial development if irrigated acreage is minimal and the potential for water conservation is negligible. Again, small landscape projects should not be unnecessarily burdened with an extensive submittal and review process.

Industrial Development

New industrial parks and complexes also may include large landscapes. Contemporary designers often use large areas of nonfunctional turfgrass and other features that use large amounts of water. Many industrial sites could vastly improve their landscape water efficiency. As with commercial development, excluding minimal landscape area from regulation may be appropriate.

Public Authorities

Public landscapes provide an opportunity to demonstrate the benefits and appeal of water-efficient landscaping. Local municipalities often are responsible for irrigating parks, gardens, street medians, and landscapes of public facilities. Public authorities can provide leadership in implementing water-efficiency requirements for their own landscapes. In addition, considerable fiscal savings can result when municipalities incur high costs for landscape irrigation.

EXCLUSIONS FROM APPLICABILITY

Landscape standards discussed in the following paragraphs may be inappropriate because of limited potential for actual water savings. Administrative capacity and cost-effectiveness need to be examined. Also, certain regulations may be inappropriate to some specialized landscape irrigation.

Landscaping Provided by Homeowners

Actual water savings simply may not justify including small residential landscapes in the range of applicability. In high-density residential developments with minimal landscaped areas, exclusion may be necessary to avoid imposing unenforceable regulation.

Botanical Gardens, Parks, and Recreational Areas

Restricting the use of certain types of plants and irrigation may conflict with the needs of some educational and recreational facilities. It may be necessary to exempt these areas from specific regulations to preserve their value to the community. In most cases, overall conservation goals will not be significantly affected. However, efficient management should be required for all sites.

Edible Crops

Water-efficiency requirements could adversely affect residential users who grow edible crops. Strict restrictions on using plants that need lots of water could preclude the planting of vegetable gardens. Edible crops may need special consideration, depending on the nature of the regulations. However, irrigation of edible crops can be subject to efficiency requirements, as with ornamental landscapes.

The consumption of water within an urban jurisdiction by commercial agriculture should be treated separately from ornamental use and is not within the scope of this publication.

EXISTING AND REHABILITATED LANDSCAPES

Landscape renovation that requires a construction permit and exceeds a set, locally determined cost also can be subject to regulation. Potential for water savings should be large enough to justify inclusion in the range of applicability.

Rehabilitated landscape should be clearly defined in the regulations. The definition needs to be determined locally, using a project cost as the criteria or local requirements for landscape construction permits.

RECLAIMED WATER

Reclaimed water is treated, recycled wastewater. The potential for using reclaimed water depends on the degree of treatment, how the irrigated area is used, and the economic benefits. Wastewater treatment can be done on-site if an industrial or commercial use generates an adequate supply or reclaimed water may be available from wastewater treatment plants. A delivery system, separate from the potable water supply, usually needs to be developed to bring reclaimed water to a site.

Use of reclaimed water is more feasible for larger sites, such as golf courses and large industrial parks, where acreage allows storage of the reclaimed water in ponds until irrigation is needed. Investment in reclaimed water systems is more cost-effective for large sites due to the high demand.

It may be necessary to apply reclaimed water at rates above the plant's minimum requirement (Ferguson 1988). Even highly treated wastewater can cause harmful salt buildup in the root zone. Increased applications, usually 5 percent to 10 percent above the normal requirements, are necessary to leach salts from the plant root zones.

In designing systems that use reclaimed water, selecting alternative, salt-tolerant species and the capacity for natural precipitation to leach salts from the soil should be considered. In addition, since the quality of reclaimed water can vary significantly, depending on the source and level of treatment, purging soil with salt-laden water may add to the problem.

Efficiency standards for irrigation with potable water may not be appropriate for reclaimed water use, since allowances must be made for leaching salts. Where wastewater is treated on-site, biological processes that purify water as it moves through soil are sometimes used as the final stage in wastewater treatment. Landscapes that use a high volume of water may be justified where they are necessary to dispose of highly treated wastewater, some of which infiltrates to the depth of the water table and may be stored as groundwater.

Where it is practical to require reclaimed water use, additional economic incentives may be provided to users. This may be done through a rate structure, connections fees, or subsidies for reclaimed water installations.

REFERENCE

Ferguson, B.K. 1988. "Using Water Effectively." In *Irrigation, Volume 3 of Handbook of Landscape Architectural Construction.* Edited by Scott Weinberg and John Mack Roberts. Landscape Architecture Foundation, Washington, D.C.

WATER-EFFICIENT LANDSCAPE GUIDELINES 23

Irrigation Design

▲ Brown patches on the sloped portion of this lawn result from less water infiltration than on the level portions. But, trying to eliminate the brown patches often results in excess water application to the level areas. Multiple irrigation cycles of short duration would improve appearance without increasing total water application. *(Courtesy City of Phoenix)*

Irregular spacing of spray heads causes uneven application of water; too much water pressure atomizes the spray, causing it to drift from the intended area; and excess water application causes runoff. The result is poor appearance of the turf and water waste. (Courtesy City of Phoenix)

The arc patterns of these single-stream rotors are out of adjustment, which results in water being applied to paved areas. In addition, the precipitation rates of these sprinkler heads are not matched when operated in various arc patterns. Single-stream rotors without matched precipitation rates should not be used together if adjusted to different arc patterns. (Courtesy Irrigation Association)

Wasteful gutter flooding results from poor irrigation management. Some communities prohibit the use of turfgrass in narrow street medians because of the all-too-common and highly visible waste. (Courtesy Central Utah Water Conservancy District)

▲ Use of water-conserving plants does not always ensure water efficiency. A full-circle fixed spray head is inappropriate in this location because the spray pattern misses the intended area. Drip irrigation is an alternative to spray irrigation for small beds planted with trees and shrubs. *(Courtesy East Bay Municipal Utility District)*

◄ Irrigation heads can be broken by vehicular traffic, maintenance crews, or by vandals. Regular irrigation system maintenance and repair are necessary to ensure system efficiency. *(Courtesy City of Phoenix)*

Landscape Design

▲ The planting design for this landscape creates a lush appearance but requires little supplemental irrigation. *(Courtesy East Bay Municipal Utility District)*

WATER-EFFICIENT LANDSCAPE GUIDELINES 27

▲ Turfgrass in most western climates requires frequent supplemental irrigation. Turfgrass that requires supplemental irrigation should be used only where it serves a function. The turfgrass in this urban plaza can be used for foot traffic and recreation. The water feature, which has a recirculating system, requires a similar amount of water as a turf area of comparable size. *(Courtesy City of Phoenix)*

The turfgrass in this corporate landscape does not serve a practical function. However, because of the symbolic value associated with large fields of turf, turfgrass often is used to project a successful ▼ corporate image. *(Courtesy Easy Bay Municipal Utility District)*

28   WATER-EFFICIENT LANDSCAPE GUIDELINES

▲ Besides requiring frequent irrigation, this typical use of turfgrass does not complement the arid natural landscape. *(Courtesy East Bay Municipal Utility District)*

If a green oasis is desired in an arid climate, water-conserving trees, shrubs, and ground covers can create an interesting design that better complements the natural ▼ landscape. *(Courtesy Northern California Turfgrass Council)*

◀ An attractive water-conserving street median reduces maintenance costs because regular mowing and irrigation are not required. A border of decomposed granite improves safety for maintenance crews. *(Courtesy East Bay Municipal Utility District)*

▲ This attractive, low-maintenance highway median complements the natural landscape and can thrive with no supplemental irrigation. *(Courtesy City of Phoenix)*

◀ A grassy berm in a narrow street median is a recipe for water waste and increased maintenance costs. Because medians usually are irrigated with spray irrigation, the narrow width results in overspray and the high precipitation rates result in runoff. In addition, regular mowing is required, and mowing equipment must be operated among vehicular traffic. *(Courtesy East Bay Municipal Utility District)*

▲ Eliminating supplemental irrigation by using native and adapted plants is the best way to ensure water efficiency. *(Courtesy East Bay Municipal Utility District)*

◀ The convenience of automated irrigation makes over-irrigation of residential landscapes more likely. *(Courtesy East Bay Municipal Utility District)*

WATER-EFFICIENT LANDSCAPE GUIDELINES 31

(Courtesy Northern California Turfgrass Council)

(Courtesy East Bay Municipal Utility District)

▲ Water-efficient landscaping does not have to be stark or dry in appearance.
◄ This landscape (left) in a hot climate results in excessive glare and heat gain. Reduced lawn area and a variety of interesting plants (above) minimize water requirements.

▲ Landscaping constructed from hard materials (sometimes called hardscape), such as paving, provides functional spaces that require no irrigation. Hardscape can enhance both residential (above) and commercial (left) ◀landscapes. *(Courtesy Northern California Turfgrass Council)*

WATER-EFFICIENT LANDSCAPE GUIDELINES

Overhead irrigation of sloped terrain should be avoided. Poor sprinkler head spacing results in brown patches, but closer spacing will increase application rate, resulting in more runoff into the gutter.
(Courtesy Irrigation Association)

▲ To reduce potential water waste, trees and shrubs on a slope can be irrigated with drip irrigation.
(Courtesy East Bay Municipal Utility District)

34   WATER-EFFICIENT LANDSCAPE GUIDELINES

▲ This industrial landscape uses reclaimed water from an on-site wastewater treatment system for ornamental ponds and for landscape irrigation. *(Courtesy East Bay Municipal Utility District)*

◄ As with this residential garden, water efficiency can be achieved without sacrificing aesthetic value. *(Courtesy East Bay Municipal Utility District)*

Landscape guidelines and regulatory programs can promote the preservation of natural landscape areas and native plant communities such as these. *(Courtesy East Bay Municipal Utility District)*

Landscape Management

▲ Damage to the pavement from over- irrigation of this grassy parking lot island will result in costly repair. *(Courtesy East Bay Municipal Utility District)*

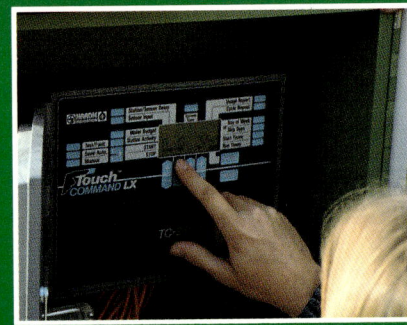

▲ Accurate programming of irrigation controllers is essential to water efficiency. An irrigation controller is programmed to operate remote-control valves. *(Courtesy East Bay Municipal Utility District)*

◀ Operating pressure can be tested in the field with a pressure gauge. *(Courtesy Irrigation Association)*

▲ Precipitation rates and uniformity of application can be calculated by analyzing the results of catch-can tests. *(Courtesy East Bay Municipal Utility District)*

◀ Meter readings can be used to survey landscape water use and detect leaks in system hardware. *(Courtesy East Bay Municipal Utility District)*

38 WATER-EFFICIENT LANDSCAPE GUIDELINES

▲ Education is essential to promoting design and management efficiency. *(Courtesy East Bay Municipal Utility District)*

◄ Landscape managers use information from weather stations to precisely determine irrigation schedules. Evapotranspiration data can be disseminated through a telephone hotline. *(Courtesy East Bay Municipal Utility District)*

4

Water Budget Approach

Before discussing the water budget approach in depth, it will be helpful to define a few terms.

DEFINITIONS

A landscape water budget, or water allowance, is the product of a formula that quantifies a volume of water for landscape irrigation. The factors that comprise a water-budget formula determine the efficiency standards for the design and management of a landscape site. Factors are derived from the estimated water needs of plants and other landscape features that use water and from measures of irrigation efficiency. Since rainfall will impact landscape water requirements only when water is needed, effective precipitation (see definition below) is used to more closely approximate the impact of rainfall.

A **water budget,** or water allowance, is a site-specific quantity of water needed to maintain an ornamental landscape's plants and other features that use water. The amount of water required for plants is combined with an irrigation efficiency factor to form a constant standard that is applied to specific sites. The area to which the budget applies is the variable that ties a budget to a specific site. Water budgets can be calculated from plans of new landscapes or from site measurements of existing landscapes.

Effective precipitation, or beneficial rainfall, is the portion of rainfall available to plants that does not exceed the plants' requirements. Some precipitation will create surface runoff or percolate below the plant root zones rather than becoming available to the plants. Of the precipitation that does become available, only the maximum amount required by the plants is considered effective or beneficial in meeting plant needs.

Evapotranspiration (ET) is a measure of water required by plants that includes water losses due to evaporation from the soil and transpiration through plant foliage. ET was developed for scheduling agricultural irrigation and can be used to determine an ornamental landscape's water requirements.

40 WATER-EFFICIENT LANDSCAPE GUIDELINES

Using a water budget approach depends on the climate and the resulting water requirements of plants. In arid climates or Mediterranean climates with periods of annual drought, ornamental landscapes can depend entirely upon supplemental irrigation for most of the year. Where precipitation is near to or exceeds ET rates, supplemental irrigation usually is not required.

Figures 4-1 and 4-2 illustrate the variation between climates in Atlanta, Ga., and Walnut Creek, Calif. In Atlanta, rainfall will meet the needs of most plants

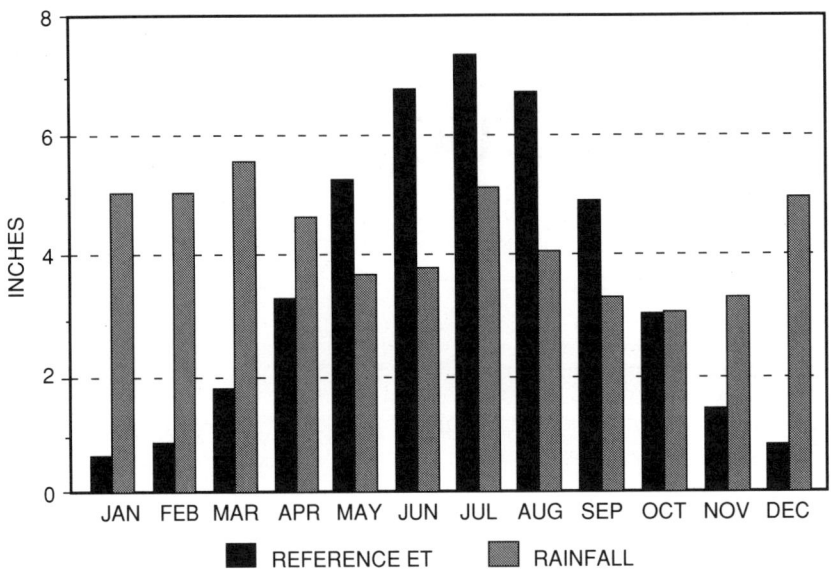

Figure 4-1 Reference ET and rainfall in Atlanta, Ga.

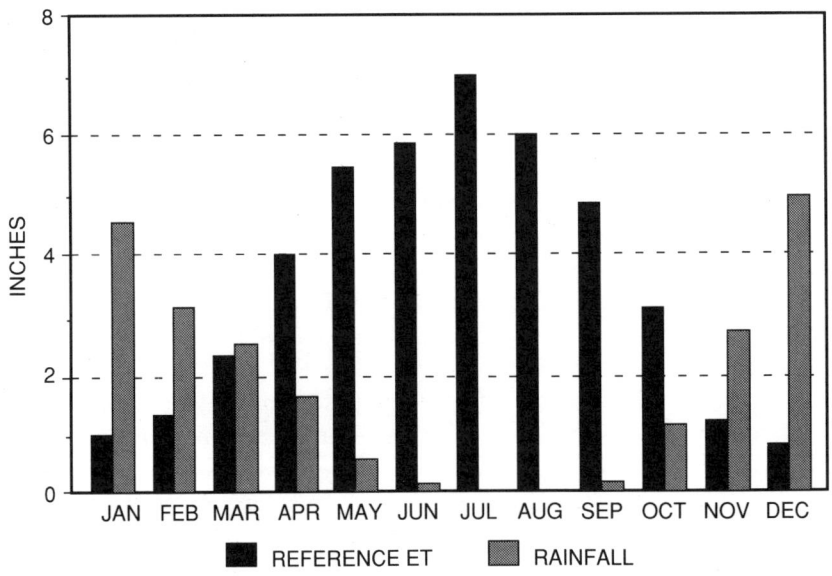

Figure 4-2 Reference ET and rainfall in Walnut Creek, Calif.

during all the months of the year, provided an average rainfall occurs. However, supplemental irrigation may be required during episodic drought; a water budget may define consumption limits during a drought.

In Walnut Creek, plants unadapted to annual summer drought will consistently require supplemental irrigation from April through October. Rainfall is not a factor during most of the irrigation season, and a water budget can be a good predictor of seasonal water requirements for plants.

Water budget formulas must account for actual water requirements of plants and acceptable levels of irrigation system hardware and management efficiency. Budgets then can function to limit the use of elements in the landscape that consume water and to provide incentive for efficient irrigation.

HISTORIC BASIS VERSUS ACTUAL LANDSCAPE WATER REQUIREMENTS

Existing landscapes also can be subjected to water budgets. The option of using prior consumption to determine appropriate usage limits is feasible where consumption has been metered and historic data is available.

Advantages to using historic data as a basis for setting limits are its availability and the elimination of a submittal and review process. Water utilities often use a percentage of prior consumption to set limits on various types of use during emergency water shortages. However, while prior consumption can be used as a stop-gap method during drought emergencies, it is neither fair nor effective in regulating long-term landscape water efficiency. That is because it effectively rewards inefficient irrigation by allowing more water to customers with historic inefficiencies and less to those who were efficient.

WATER BUDGET FACTORS

Water budget factors include: area, water requirements of plants and water features, irrigation efficiency, and conversion factors to obtain the desired unit of measure. Each factor is examined below.

Area

The following paragraphs explain how to define and determine the area to be regulated.

Defining the area. Appropriate efficiency standards depend on a clear definition of the surface area to be regulated. Area is expressed in acres, or more commonly, square feet. Area can be defined as:

- **site area**—the total area of a site, including building footprints, roadways, and parking areas (Figure 4-3)
- **landscape area**—the combination of planted areas, water features, hardscape (landscaping constructed from nonliving materials, such as concrete, tile, and lumber), and nonirrigated planted or undeveloped areas (Figure 4-4)
- **irrigated area**—planted areas requiring supplemental irrigation (Figure 4-5)

New landscapes. Irrigated area taken from a final plan can be used as the area factor in a water budget formula; the efficiency standard inherent in the formula applies to the irrigated area. However, in the predesign stages of a project, the

42 WATER-EFFICIENT LANDSCAPE GUIDELINES

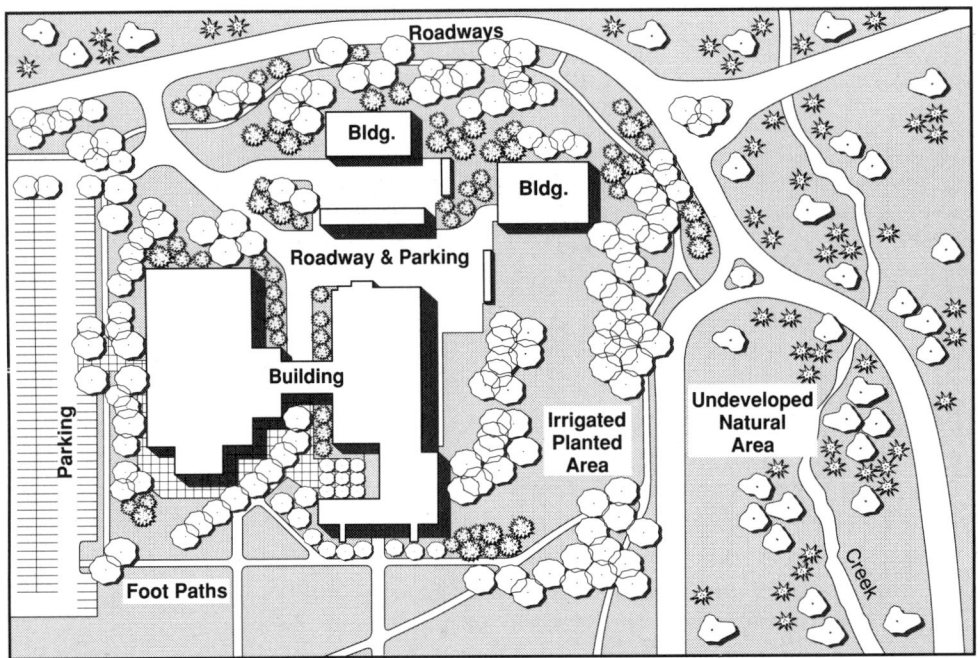

NOTE: Site area includes irrigated planted areas, nonirrigated planted areas, undeveloped natural areas, roadways, parking areas, and building footprints.

Figure 4-3 Site area

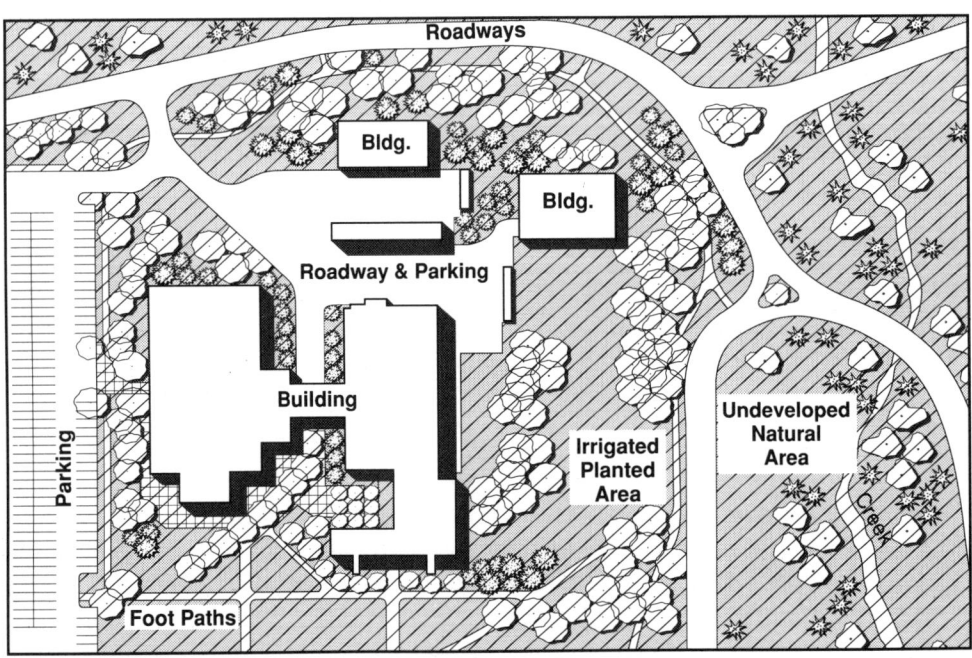

NOTE: Landscape area includes both irrigated and nonirrigated planted areas, undeveloped natural areas, water features, and hardscape, such as paths, decks, and patios.

Figure 4-4 Landscape area

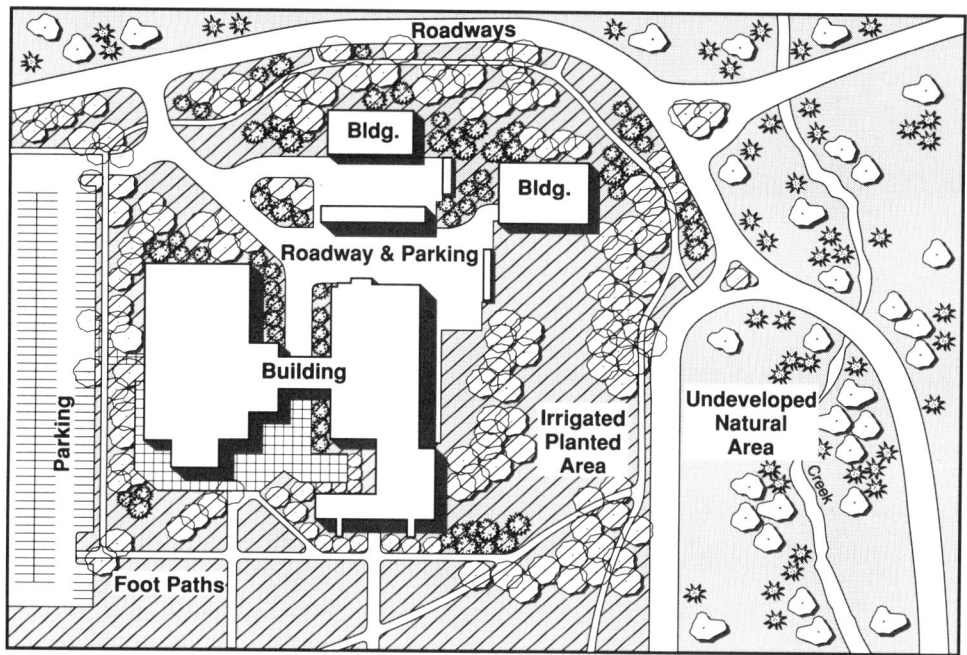

NOTE: Irrigated area includes planted areas under irrigation.

Figure 4-5 Irrigated area

irrigated area is not known. If the building footprints, roadway, and parking areas have been defined, landscaped area can be used to calculate a water budget.

A water budget based on landscape area can function as a design guideline. The resulting budget should be strict enough to limit the amount of irrigated area and encourage water-thrifty design of irrigated and nonirrigated areas.

Nonirrigated areas can include undeveloped or unlandscaped natural areas. Preservation of such areas can be encouraged by establishing water-budget limits that cause designers to concentrate irrigation in smaller, selected areas of a site. Thus, the designer can use a site's water budget as a design tool by comparing the water requirements of proposed irrigated areas to the water budget based on landscape area.

Existing sites. Irrigated area should be used to calculate water budgets for existing sites because such budgets establish management standards for all areas under irrigation without requiring extensive changes in design or loss of existing plantings. A water budget based on landscape area may allocate water for nonirrigated areas and allow inefficiency in the irrigated areas.

Determining area. Landscape and irrigated areas can be determined from planting or irrigation plans, aerial photos of large sites, or on-site surveys. Use of planting and irrigation plans is preferred because they are drawings from a set of landscape plans that contain information relevant to water efficiency.

Irrigation and planting plans drawn to scale should clearly indicate the scale. A bar scale should be included to ensure the drawings are not reductions or enlargements of the originals. Area can be measured using an architect's or engineer's scale or a planimeter. Calculations of area from plans, called area takeoffs, are performed by civil engineers, landscape architects, and contractors for design and estimating purposes. Detailed, accurate area takeoffs may take considerable time to perform but often are done as a normal part of design and cost estimating processes.

Two-dimensional plans of sloped sites do not exactly represent a site's actual surface area. Depending on the degree of the slope (determined from a grading plan or site survey), the area extracted from the plan can be corrected by a factor that accounts for the slope. However, corrections for sloped areas usually are not factored in area calculation because site dimensions, including dimensions in the legal description of a property, are typically measured in plan view. (A plan view is a two-dimensional representation of the ground plane as viewed from overhead.)

Where landscape and irrigation plans are not available for larger sites, aerial photo prints can be obtained from the local engineering or planning department or be purchased from a local aerial photo company. The area of a site can be measured from photo prints, as with landscape and irrigation plans, provided the scale is accurately and clearly stated. Area also can be determined through an on-site survey done by a qualified landscape professional or surveyor.

Landscape Water Requirements

ET, a measure of water required by plants, usually is expressed in inches but can be converted to gallons or units of 100 cubic feet (Ccf) as discussed later. Reference evapotranspiration "... is the ET of a broad expanse of well-watered, 4-to-6-inch-tall cool-season grass" (Snyder, Harivandi, and Lanini 1991). A regularly mowed, cool-season turfgrass will not require 100 percent of reference ET (ET_O), and many trees and shrubs will require substantially less than ET_O. As shown below, ET_O is adjusted by a crop coefficient (K_c) to quantify an individual species' water requirement (ET_c).

$$ET_c = ET_O \times K_c$$

Where:

ET_c = ET rate of a specific crop
ET_O = reference evapotranspiration
K_c = crop coefficient

ET_c rates have been established for turfgrass and agricultural crops. K_c values for most ornamental plants have not been developed by field measurements. In addition to the lack of empirical data for ornamental plant species, use of K_c factors is complicated because ornamental landscapes are not monocultures where one K_c value applies to an entire irrigated area.

However, it is possible, based on K_c factors for turfgrass and agricultural crops and known water needs for ornamental plants, to determine appropriate values for various types of ornamental landscape planting (Costello, Matheny, and Clark 1991). In an arid, mild-winter climate, the water requirement for a regularly mowed, cool-season turfgrass is 80 percent of ET_O because of the foliage's reduced surface area. Cool-season turfgrass defines the high end of water requirements for ornamental plants, and for the purpose of the following discussion, high, medium, and low categories are defined.

High-water-use plants are plants with requirements of 50 to 80 percent of ET_O. Some trees, shrubs, and perennials will have high requirements similar to turfgrass.

Medium-water-use plants are plants requiring 30 to 50 percent of ET_O. Low-water-use plants require less than 30 percent of ET_O. Plants native to a region or from similar climates are naturally adapted, so established plantings may require no supplemental irrigation.

Other factors, such as plant maturity, planting combinations, density, microclimates, and plant rooting depth, also will contribute to actual water requirements. Not all of these factors can be quantified in the design stage of a project, nor are they necessary to determine an appropriate water budget for a given landscape and climate. Rather, these additional factors become more important when determining efficient irrigation schedules and are discussed in "Scheduling Standards," chapter 5.

Historical ET$_O$ data. Historical ET$_O$ data usually are based on average climate data. Monthly ET$_O$ tends to vary little from year to year and is a reliable measure of plants' water requirements. Precipitation, however, will vary widely; where plant material largely depends on natural precipitation, supplemental irrigation may be required only during periods of below-normal rainfall. However, these periods may concur with low water supply and the need to restrict use.

In arid and semiarid climates, historic ET$_O$ will serve as an adequate basis for determining water requirements for plants since irrigated landscapes rely almost entirely on regular irrigation during the growing season. Most plants, especially water-conserving plants, usually can survive periods of stress due to drought and will survive on irrigation schedules based on historical averages (Figure 4-6).

Corrections for rainfall. Because effective precipitation is not predictable, subtracting average historic rainfall from a water budget may result in water allotments well below actual ET rates during periods of below-average rainfall. Conversely, failing to subtract effective precipitation rates from a water budget will weaken the efficiency standard and the incentive for efficient management.

Only a portion of total rainfall can be considered effective in sustaining plants. Effective precipitation, or beneficial rainfall, is the portion of rainfall that does not exceed the depletion of water from the soil through ET during a given period. For example, in Mediterranean climates, with winter rain and annual summer drought, most precipitation occurs when ET rates are lowest. This minimizes the beneficial rainfall and the need to adjust for rainfall.

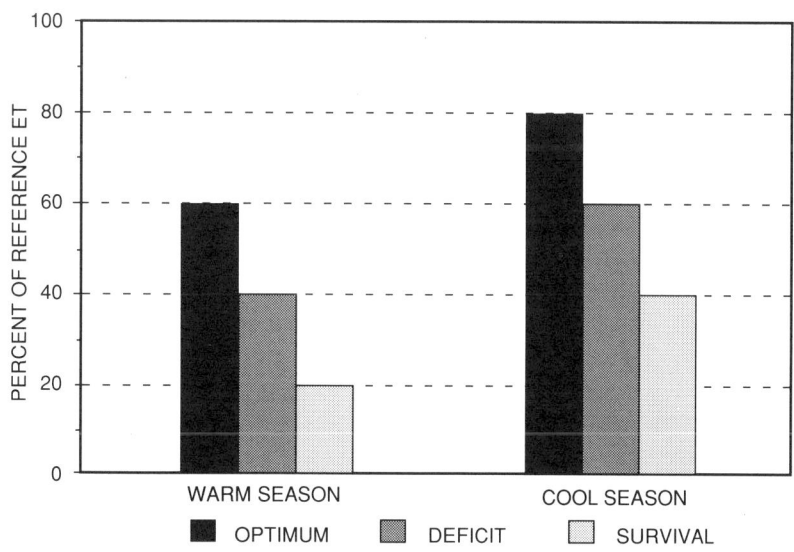

Source: *Gibeault et al. (1991).*

Figure 4-6 Turfgrass requirements at optimum, deficit, and survival levels or irrigation

Where precipitation occurs throughout the year, beneficial rainfall has a greater impact on actual water requirements and may obviate the need for supplemental irrigation.

Empirical basis for water budgets. While the water budget approach has been criticized for a lack of empirical data on which to base water budget formulas, adequate data on water requirements for plants is available to determine appropriate budgets. ET rates of turfgrass have been established through field testing, and water requirements of trees and shrubs are known through practical horticultural experience. This practical knowledge is well documented in plant books and encyclopedias.

The value of more refined data is limited for two reasons:

1. Ornamental landscapes are comprised of mixed plantings affected by a variety of mitigating variables.

2. Small variations in water requirements for plants do not significantly impact a total water budget.

Nevertheless, more accurate data is being developed that will be valuable in generating irrigation schedules for newly designed and existing landscapes. One such research project tapped the knowledge of experienced landscape professionals and applied a consistent evaluation method to generate reasonable assessments of the water requirements of many plant species (Costello and Jones 1992).

Irrigation Efficiency Standards

In addition to landscape water requirements, irrigation inefficiencies contribute to water consumption in landscapes. Standards for irrigation efficiency can be factored into a water budget. Irrigation efficiency (IE) is the amount of water beneficially applied, divided by the total amount of water applied. As a percentage:

$$IE = \frac{\text{amount of water beneficially applied}}{\text{total amount of water applied}} \times 100$$

Two factors, system efficiency and management efficiency, contribute to IE.

Hardware efficiency. Hardware efficiency, or system efficiency, for sprinkler irrigation is calculated as a measure of the uniformity of application. Precipitation from irrigation heads does not deliver the same amount of water to all parts of an irrigated area. Some areas will receive more water than required in order for the whole site to receive an adequate amount.

A statistical measure of distribution uniformity (DU) is determined by analyzing the results of "catch-can" tests. Determining DU is an established landscape water-auditing procedure. Assuming no overspray or runoff occurs, DU is an acceptable measure of system efficiency for spray irrigation. Uniformity of 65 percent to 75 percent is an industry standard.

The shape and dimensions of an irrigated area will affect the level of uniformity that can be achieved. Higher uniformity will be achieved more easily with new installations specially designed to maximize the effectiveness of spray pattern uniformity. The same level of efficiency will be more difficult to achieve in retrofitting poorly designed installations.

Hardware efficiency for drip irrigation is more difficult to calculate because uniform delivery of water to plant roots relies on lateral, subsurface movement of water. Water dispensed at the surface creates a wetting pattern that will vary with soil type, emitter placement, flow rates, and duration of irrigation cycles. Furthermore, uniform wetting of the root zone is not as critical for trees and shrubs as for continuously rooting ground covers. System efficiency for a properly designed drip irrigation is 80 percent to 95 percent depending on climate and soil variations.

Management efficiency. Management efficiency is a function of system maintenance and scheduling. When uncorrected, malfunction of system components can account for a large part of water waste. Improper scheduling that does not respond to seasonal changes and plants' water requirements compounds water waste. But even with careful attention to maintenance and scheduling, some water will not be beneficially applied.

While landscape managers should strive for 100 percent management efficiency, it is reasonable to allow some margin of inefficiency when establishing water budgets. For example, some consumption may occur during routine testing and flushing procedures that are necessary to ensure proper function. And, not all leaks can be prevented or detected immediately.

Irrigation should be scheduled to cause wetting of the soil to the depth of plant root zones. Some moisture, however, can be expected to infiltrate soil where it is not available to plant roots. As for frequency, irrigation should be scheduled to replace depletion of soil capacity by ET. Unless irrigation control is directly tied to real-time ET data, some error in scheduling frequency may cause applications to exceed the soil's moisture-holding capacity.

Quantifying an acceptable standard for management efficiency is difficult because field testing and research that specifically address this issue are unavailable. Research also has indicated that turfgrass can be effectively managed below ET_c levels (Gibeault et al. 1991). This ability of plants to survive below their respective ET_c levels may negate the need to factor for normal scheduling errors. A management efficiency factor of 80 percent to 90 percent will allow for most normal errors in scheduling and maintenance.

Irrigation efficiency. The product of hardware efficiency and management efficiency yields a factor that can be used to adjust ET_O rates for a landscape. This factor is called the irrigation efficiency (IE):

$$IE = (hardware\ efficiency)(management\ efficiency)$$

From the above standards, the following IE is obtained for sprinkler irrigation:

$$IE\ for\ sprinkler\ irrigation = (0.75)(0.9)$$

$$IE_{(sprinkler)} = 0.675$$

Some landscape regulations call for all trees and shrubs to use drip rather than sprinkler irrigation, because the efficiency of drip irrigation is assumed to be higher than for spray irrigation. Note, though, that in the absence of management efficiency, the differences found in system efficiency are secondary. That is because a properly managed sprinkler system can be more efficient than a poorly managed drip system. Conditions at each site and type of plants to be irrigated determine the appropriateness of sprinkler irrigation. Figures 4-7, 4-8, and 4-9 illustrate the combined effect of irrigation efficiency and plant water requirements on the total amount of water applied.

Hardware efficiency of 90 percent, which is obtainable with drip irrigation, will yield a higher IE value when combined with a management factor:

$$IE_{(drip)} = (0.9)(0.9)$$

$$IE_{(drip)} = (0.81)$$

48   WATER-EFFICIENT LANDSCAPE GUIDELINES

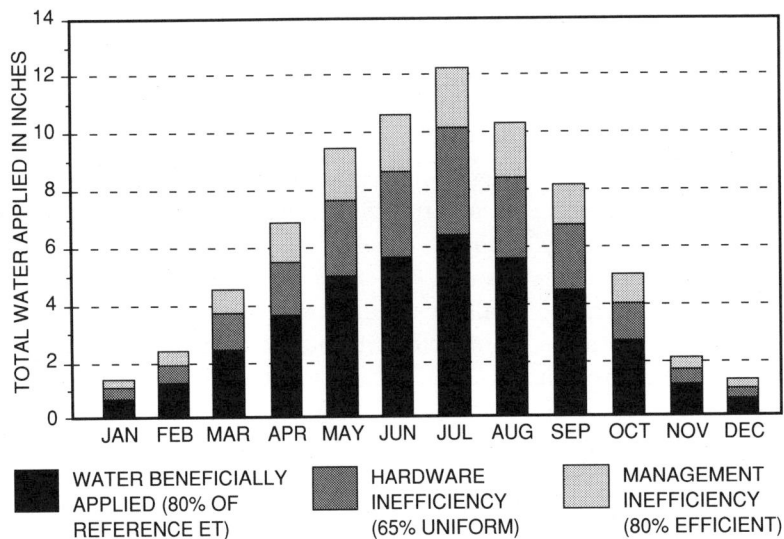

Figure 4-7 Sprinkler irrigation efficiency for bluegrass in average California climate conditions

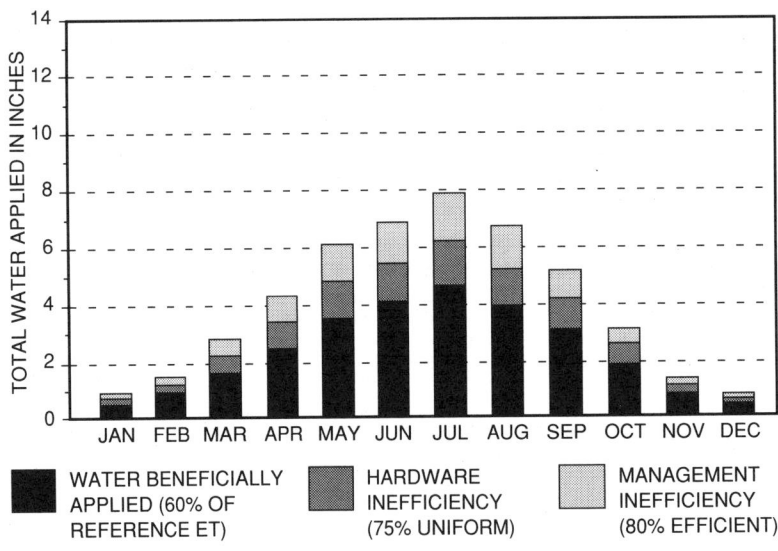

Figure 4-8 Sprinkler irrigation efficiency for water-conserving turfgrass and 75 percent uniformity in average California climate conditions

Conversion Factors

ET expressed in inches is primarily useful for scheduling sprinkler irrigation where application rates are measured in inches per hour. ET expressed in inches represents a quantity of water lost from the soil that can be replaced accurately if the precipitation rate and a system's level of uniformity are known. Because ET is usually expressed in inches, the quantity must be converted to gallons or billing units.

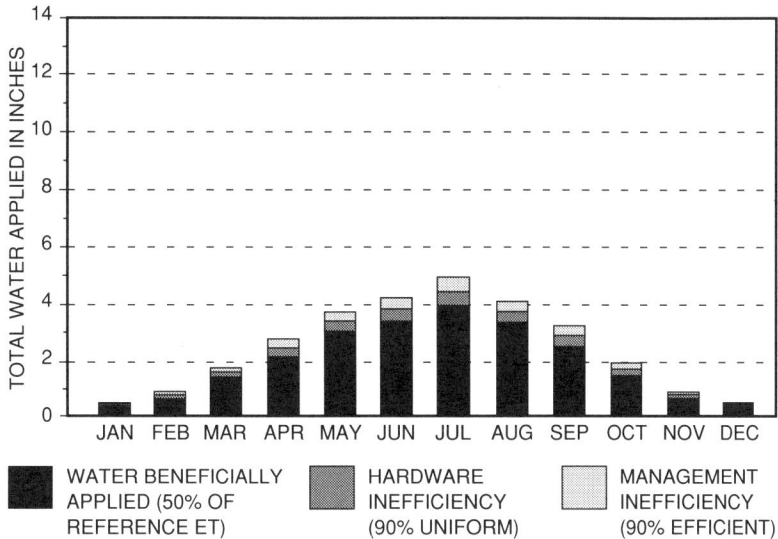

Figure 4-9 Drip irrigation efficiency for moderate plant water requirement in average California climate conditions

To generate a water budget for a parcel of land, a factor of 0.623 will convert ET for that parcel of land to gallons. It takes 0.623 gallons to create a 1-in.-deep layer of water per square foot (square foot inch). The product of the following factors, then, will yield gallons:

(square foot of area) (inches of ET) (0.623 gal/square foot inch) = gallons

A commonly used billing unit is Ccf, or 748 gal. Conversion of a water budget to Ccf may be desirable or necessary if a budget will be reviewed and enforced through a rate structure established by a water purveyor.

The factor used to convert the product of the area in square feet and ET in inches to Ccf is 0.00083, as shown below:

(square foot of area) (inches of ET) (0.00083 Ccf/square foot inch) = Ccf

WATER BUDGET FORMULAS

Next, this book will show how to derive a basic formula for establishing a water budget, then discuss several variations of this formula.

Simple Water Budget Formula

This simple water budget formula will approximate yearly water requirements for a specific landscape:

(annual ET_O) (area in square feet) (conversion factor) = annual budget (Eq 4-1)

This formula does not account for IE (which will increase the budget) or limits on high-water-use landscape elements (which will decrease the budget). In addition, to reflect the seasonal variation of water requirements, a yearly allowance should be distributed throughout the year to reflect average monthly water requirements.

Adjustment Factor

Adjustment factors can be introduced to refine this formula. As discussed above, ET_O is a reference for quantifying a landscape's water requirements. One adjustment factor to consider is a plant factor, which is a percentage of ET_O required by a water-conserving landscape.

An average plant factor applied to the landscape area will effectively limit the use of high-water-use plants and water features. Water features that recycle water, such as fountains, ponds, and swimming pools, have similar requirements to high-water-use plants and should be given similar value in establishing a water budget. Restricting the design and size of water features will reduce consumption. Another element to consider is that fountains with high-volume spray jets will have more evaporative loss and losses due to wind drift.

A formula for determining an average plant factor should be based on acceptable standards for plants' water requirements rather than an arbitrary reduction of ET_O. Checklist approaches often define percent limits on uses of certain types of plants. Table 4-1 illustrates the effect an average plant factor (APF) may have on selecting plants.

A similar APF is shown in Table 4-2. By including nonirrigated areas, larger areas of turfgrass can be accommodated. For larger sites, inclusion of nonirrigated areas may result in the preservation of native landscapes and ecosystems that exist on a development site.

The APFs in Tables 4-1 and 4-2 represent a moderate standard for landscape water requirements. If high water use were allowed for the whole landscape area, an average factor of 0.8 would result. This would not represent an efficient design

Table 4-1 Plant selection resulting in a moderate average plant factor

Plant Type by Water Requirement	Plant Factor Average ET_c %	Percent of Landscape Area	Weighted Factor* %
Turfgrass	80	25	20
High water use	80	10	8
Medium water use	50	20	10
Low water use	30	45	13.5
Average Plant Factor†			51.5

*(Plant factor as a percentage) × (Percent of landscaped area).
†Sum of weighted factors.

Table 4-2 Use of nonirrigated areas to accommodate high-water-use elements

Plant Type by Water Requirement	Plant Factor Average ET_c %	Percent of Landscape Area	Weighted Factor* %
Turfgrass	80	60	48
High water use	80	5	4
Nonirrigated	0	35	0
Average Plant Factor†			52

*(Plant factor as a percentage) × (Percent of landscaped area).
†Sum of weighted factors.

standard for two reasons: alternative plant material can be used without losing form or function and most plants will thrive under irrigation that is lower than established ET_C rates.

A strict standard would require all plants to be low water users resulting in an average plant factor of 0.3. In this case, either high-water-use plants or total planted area would be limited by a 0.3 plant factor.

Plant factors should be based on local conditions and water-conservation goals. Establishing a plant factor allows some design flexibility as long as the APF is not exceeded.

Determining adjustment factor. An APF for a landscape can be divided by the IE to yield an ET_O adjustment factor. This adjustment factor (AF) represents a combined standard for plant water requirements and IE, as shown below.

$$AF = APF/IE$$

Where:

AF = ET_O adjustment factor
APF = average plant factor
IE = irrigation efficiency

The AF is critical to establishing water-efficiency standards. It will vary depending on the landscape's water requirements, as represented by the APF. From the previous discussion, an APF of 0.8 for all high-water-use plants, when divided by the IE, results in an AF of 1.2 (0.8/0.65 = 1.2). The IE stated before has been rounded to 0.65. As a percentage, the AF is 120 percent of ET_O.

A moderate plant factor of 0.6 will yield about 100 percent of ET_O (0.6/0.65 = 0.92). Where a moderate plant factor is an adequate standard for landscape water requirements, an AF is superfluous. ET_O rates then can be used with no adjustment. Equation 4-1, the simple water budget formula, then should be used to determine the annual water budget.

A strict plant factor of 0.3 results in a reduction of ET_O (0.3/0.65 = 0.46). This represents 46 percent of ET_O. Water-conserving plants (with low APFs) under drip irrigation (with high IE) could be sustained within these parameters. Concentrated areas of landscape that use high volumes of water also may be accommodated by preserving nonirrigated landscape areas.

Adjusted Annual Water Budget Formula

A more complete water budget formula will incorporate the AF:

annual water budget = (local annual ET_O) (adjustment factor)

(landscape area) (conversion factor)   (Eq 4-2)

An annual water budget needs to be distributed throughout the year to reflect the seasonal variation in a landscape's water requirements. To establish monthly or bimonthly water allotments, monthly ET_O can be used:

monthly water budget = (local monthly ET_O) (adjustment factor)

(landscape area) (conversion factor)   (Eq 4-3)

Other factors such as plant grouping and density, microclimates, and beneficial rainfall also affect actual water requirements for a landscape. This type of detailed analysis is not possible without a design or existing conditions from which to derive these additional factors. Nevertheless, the above formula adequately defines a water budget based on established data and can be applied to both existing and proposed landscapes.

For projects in the planning stages, landscape area combined with a moderate-to-low ET_O AF will restrict the design of landscape elements that consume a high volume of water. This definition of landscape area is necessary because the size of the irrigated area is unknown. A design's compliance to a water budget can be determined by analyzing design elements and IE. (Analysis is performed by determining the square footage of design elements with varying water requirements. The method is described in chapter 6.)

For existing landscapes, the irrigated area should be the variable factor in a water budget formula. Some sites may have irrigated areas with plantings that do not require supplemental irrigation. These areas should be identified and excluded from the total irrigated area. An AF that is a product of acceptable IE and a moderate to high APF will represent an adequate water-efficiency standard for all areas requiring irrigation.

REFERENCES

Costello, L.R. & Jones, K.S. 1992. WUCOLS Project: Water Use Classification Project. University of California Cooperative Extension, San Francisco and San Mateo County Office, Half Moon Bay, Calif.

Costello, L.R.; Matheny, N.P.; & Clark, J.R. 1991. Estimating Water Requirements of Landscape Plantings, The Landscape Coefficient Method. University of California Cooperative Extension Leaflet 21493, Division of Agriculture and Natural Resources, Oakland, Calif. (July 1991).

Gibeault, V.A. et al. 1991. Managing Turfgrasses During Drought. University of California Cooperative Extension Leaflet 21499, Division of Agriculture and Natural Resources, Oakland, Calif. (June 1991).

Snyder, R.L.; Harivandi, A.M.; & Lanini, B.J. 1991. Turfgrass Evapotranspiration Map Central Coast of California. University of California Cooperative Extension Leaflet 21491, Division of Agriculture and Natural Resources, Oakland, Calif. (March 1991).

5

Checklist Approach

The checklist approach to guidelines and standards addresses specific criteria for water efficiency in landscape design and management. This approach is necessary when water consumption cannot be monitored because irrigation use is not metered or because an enforcing agency does not have access to consumption information. This chapter presents an overview of the various issues related to water-efficient design and management. The following discussion may be used to generate guidelines or to establish regulatory standards suited to a given region.

LANDSCAPE DESIGN STANDARDS

A checklist approach to water-efficient design is most effective when guidelines or standards can be quantified. Specific, quantitative standards are preferable to general conceptual requirements, which are more useful as educational tools and guidelines. Landscape design standards related to functional design, soil improvement, grading, plant selection and grouping, and turfgrass limitations are discussed below.

Designing for Function

This chapter will examine three areas of functional design: functionally required areas, water-efficient landscape form, and materials that use little water.

Functionally required areas. Functionally required areas are the portions of an ornamental landscape intended to serve a specific use, such as pedestrian traffic, sports, or recreational activities. By limiting high-water-use elements to functionally required areas, a landscape's water requirements can be reduced. Strict limitations may infringe on the designer's flexibility because water use in a landscape does not always support a specific function or activity. Examples of functional design requirements include limitation of turfgrass to functionally required areas (also called practical turf areas) and forbidding nonfunctional water features such as ornamental ponds and fountains.

The aesthetics and symbolic value of broad expanses of turfgrass and massive water features define many traditional landscapes. Such landscapes define an image for residential, commercial, and industrial sites and are in demand. However, where extensive supplemental irrigation is required to support this, aesthetic, attractive water-efficient alternatives should be promoted.

Water-efficient landscape form. Some landscape designs are irrigated more easily than others. Neat, regular forms lend themselves to more efficient irrigation but may create monotony in a landscape. Freeflowing or irregular designs may be more interesting but they are more difficult to irrigate efficiently.

Although it is nearly impossible to regulate an efficient design, certain landscape form should be avoided in irrigated areas or the irrigation method selected needs to accommodate the design.

Accurate, uniform delivery of water by sprinkler systems is difficult to achieve in irregular, narrow, or odd-shaped areas. Sprinkler irrigation heads are designed to apply water in arc patterns of full or part circles. Rectangular spray patterns also are available but are less uniform than arc patterns.

Irrigated areas that do not conform to available spray patterns have lower uniformity of application, resulting in delivery of excess water to some areas to ensure adequate irrigation of the entire area.

Sprinkler irrigation heads are designed to deliver water over a specified radius. When possible, dimensions of irrigated areas should correspond to the radius of the irrigation heads to be used. For example, 15 ft is a common radius for spray heads. A lawn area irrigated by 15-ft heads should have dimensions that are multiples of the radius, such as 15 ft × 30 ft or 30 ft × 75 ft. Some variation in spacing from the head radius is acceptable to compensate for irregularity in site dimensions. That is because adequate delivery can be achieved with spacing at 1.1 to 1.2 times the head radius.

Overspray, the application of water outside the intended area, is also a problem in narrow or irregular areas. The width of an area irrigated by spray irrigation should be limited to the spray pattern radius of the irrigation heads to be used. Currently, spray heads will allow efficient irrigation at a minimum width of about 8 ft, as long as the area is level and not subject to consistently windy conditions.

Low-water-use materials. In many cases, alternative low-water-use plants and smaller water features can be substituted without compromising appearance. Paving and other treatments of the ground plane that do not involve plants can be used to limit irrigated areas. Plants adapted to existing climatic conditions may be irrigated initially so they may become established. Later they can be placed on a reduced irrigation schedule or not irrigated at all.

Restricting the use of high-water-use plants creates the potential for efficiency but does not ensure efficient landscape and irrigation management. Improper scheduling, inefficient management practices, and poor design of irrigation systems account for much of the water waste in landscapes.

Soil Testing and Soil Improvement

Knowledge of soil types at a given site is important to plant selection and efficient irrigation design. And, construction of large projects will require soils reports for structural engineering purposes. A report indicating soil type, soil depth, uniformity, infiltration rates, and pH should be required to verify that irrigation and plant selection are appropriate for the site.

Soil improvement often involves amending the soil according to recommendations in the soil report to improve water-holding capacity of sandy soils and drainage of heavy soils. Poor soils require careful irrigation management, and soil improvement will facilitate beneficial water applications.

Long-term effects of soil improvement through amendments, both organic and inorganic, are somewhat controversial. Some horticulturists recommend minimal disturbance of soil and selecting plants adapted to the site's soil conditions.

Soils reports will help identify soil conditions that would adversely affect water efficiency. Depending on existing soil conditions, soil improvement may not significantly affect water efficiency of irrigated areas. Because the need for soil improvement depends on individual conditions at each site, standardized criteria for soil improvement may not apply to an entire region.

Grading

Design of landscape contours is primarily a function of accommodating a site's uses and providing adequate storm drainage. Site grading and drainage can improve efficiency of irrigated areas, especially for sloped sites. Preliminary grading design of large sites usually is prepared by licensed civil engineers or landscape architects, who are part of a larger design team including architects and irrigation designers.

Traditionally, water efficiency for irrigated areas is not considered in preliminary design stages. Regulation of grading design would affect the preliminary design and the coordination of efforts by members of a design team. Grading design can impact water efficiency through minimizing slopes for irrigated areas, preserving existing top soil, and design for retention of storm drainage on-site.

Minimizing slopes in irrigated areas reduces runoff of applied water, especially from spray irrigation. Application rates of spray irrigation typically exceed the rate at which the soil can accept moisture. Irrigation of steep slopes combined with high application rates quickly generates runoff. Shortening irrigation cycles reduces runoff where slopes are shallow but does not significantly compensate for the sudden generation of runoff on steeper slopes.

Table 5-1 provides infiltration rates for various soil types and slopes.

Infiltration rates depend on a number of factors, including soil type, percentage of slope, and coverage of the ground plane by plant material and mulches. Some landscape ordinances require application rates lower than infiltration rates. This is

Table 5-1 Infiltration rate chart

Soil Texture, Type	Percentage of Slope				
	0–4%	5–8%	8–12%	12–16%	Over 16%
	Infiltration Rate (IR)* *in./h*				
Coarse sand	1.25	1.00	0.75	0.50	0.31
Medium sand	1.06	0.85	0.64	0.42	0.27
Fine sand	0.94	0.75	0.56	0.38	0.24
Loamy sand	0.88	0.70	0.53	0.35	0.22
Sandy loam	0.75	0.60	0.45	0.30	0.19
Fine sandy loam	0.63	0.50	0.38	0.25	0.16
Very fine sandy loam	0.59	0.47	0.35	0.24	0.15
Loam	0.54	0.43	0.33	0.22	0.14
Silt loam	0.50	0.40	0.30	0.20	0.13
Silt	0.44	0.35	0.26	0.18	0.11
Sandy clay	0.31	0.25	0.19	0.12	0.08
Clay loam	0.25	0.20	0.15	0.10	0.06
Silty clay	0.19	0.15	0.11	0.08	0.05
Clay	0.13	0.10	0.08	0.05	0.03

Source: Toro Company (1986).

*Assumes ground plane is covered. Figures will decrease with time and percentage of cover.

not always practical, nor is it easily quantified due to the variety of conditions that affect infiltration.

However, limits on the percentage of slope in irrigated areas will reduce wasteful runoff. Restricting overhead irrigation to areas with slopes of 16 percent or less, where infiltration rates vary from 0.31 in./h to 0.03 in./h, will minimize runoff when combined with careful scheduling. Restricting overhead irrigation to slopes of less than 10 percent is preferable, since precipitation rates can be compared more easily with infiltration rates.

Restricting use of overhead irrigation on slopes also may restrict the use of continuously rooting plants, such as turfgrass and some types of ground cover that require sprinkler or spray irrigation. As an alternative to spray irrigation, subsurface drip irrigation is becoming more feasible for such applications.

Grading should be compatible with plant selection and irrigation types. Because of local variations in soil types and terrain, standards for irrigation on slopes should be determined locally.

Preserving existing topsoil can contribute to water efficiency by retaining its superior moisture-holding capacity. Grading operations may involve removing topsoil from a site to obtain a finished grade specified in a grading plan. But topsoil is more desirable as a growing medium than subsoil. Loss of this superior soil can result in exposing hardpan, with lower infiltration rates. Thus, topsoil should be stockpiled on-site and replaced during final grading.

Retaining storm drainage on-site can reduce a landscape's water requirements by increasing infiltration and reusing the storm water for irrigation. When natural landscapes are replaced with hard surfaces, such as roofs and parking lots, runoff rates dramatically increase the demands on storm-drainage and wastewater treatment systems.

Grading can create shallow detention basins that temporarily hold water on-site for longer periods, thus increasing infiltration. Infiltration allows water to be reused by storing water in accessible parts of the landscape. "Water harvesting is a design for capturing and using storm water runoff on-site" (Ferguson and Debo 1990). Retention basins and cisterns can hold water on-site for later reuse in irrigation.

Potential design for on-site management of storm water varies with each site but may be a viable solution where adequate acreage exists. Therefore, it is not practical to require all sites to improve efficiency through use of storm water. However, incentives through regulations can encourage creative solutions to managing storm water that would otherwise place demands on wastewater systems. Where water efficiency is improved through infiltration or water harvesting, allowances could be made for high-water-use plants and water features.

Plant Selection and Grouping

Regulating plant selection by mandating the use of low-water-use plants is intended to reduce a landscape's water demand. Using water-conserving plants reduces consumption if the irrigation system is properly designed and managed. To regulate plant selections, many commercially available plants must have their water requirements defined when grown under local field conditions. Promoting the selection of low-water-use plants through regulatory and educational programs will motivate the plant-production and nursery businesses to increase their availability.

Many landscape architects/designers may react negatively to regulating plant selection. Regulation reduces the designer's flexibility and may inhibit use of untried plant species. Because of the lack of empirical data on ornamental plants' water requirements, disagreement can surface over actual needs.

While some landscape designers have in-depth knowledge of plants, many rely on limited lists of dependable plants. However, these lists do not address water efficiency and standards for plant selection and grouping. Standards can educate designers to new possibilities and improve professional practice.

Defining water requirements for plants. Regulating use of plant species requires defining water requirements for a wide variety of ornamental plants. As discussed above, empirical data for most ornamental species have not been developed through actual field measurements. But a body of knowledge exists that defines water requirements for plants based on practical horticultural experience.

Lists of plants commonly used within a region can be generated to categorize species by water requirements (Costello and Jones 1992). Any such list must be flexible to allow for additions and revisions based on subsequent research. Microclimates also can impact actual water requirements, and the same species may have dramatically different requirements given different microclimates. Another disadvantage is that the uncertainties involved in generating plant lists by category of water use can encumber the plan review process unless an extensive data base of commercially available species is developed and provided to designers.

Use of plants adapted to the local climate also can contribute to reduced water demand. Regulation may require that a percentage of selected plants be adaptable to existing climatic conditions. Where this is required, a list of native and adapted plants also should be developed and made available to designers.

Grouping plants by water requirements. Because grouping plants by water requirements is a fundamental principle of water-efficient landscape design, it should be specified in a regulatory approach. Plants with similar water requirements can be irrigated together and placed on the same watering schedule.

This is achieved by valving (or operating an irrigation circuit by a programmable, automatic valve) distinct plant groups. If plants that have different water requirements are grouped, separate irrigation circuits cannot be designed for similar plants. Thus, some plants would be over-irrigated. Grouping plants by water requirement also is known as hydrozoning and is widely accepted as an essential component of water-efficient landscape design. To define acceptable plant groupings, an accepted list of locally available ornamental plants categorized by water requirements is necessary.

Turfgrass Limitations

Limitations on turfgrass areas are most appropriate in arid climates where regular irrigation is necessary during long growing seasons. And, limiting use of turfgrass may reduce demand wherever supplemental irrigation is necessary. "Within the traditional landscape, turfgrass has received the major portion of the total landscape irrigation" (Welsh 1991).

Turfgrass also requires uniform delivery of irrigation water to sustain optimum appearance. With current sprinkler-irrigation technology, it is impossible to achieve 100 percent uniformity. Typically, some areas of a lawn will receive more water than necessary to ensure adequate delivery to all parts.

Reduction of turf areas as a percentage of the total landscape area has been advocated as a solution in landscape guidelines and mandated by ordinance. But strict limits on turf areas may unfairly single out one type of plant from other high-water-use plants. An inflexible limit on turfgrass does not address the real problem of inefficient management, nor does it preserve the benefits turfgrass provides.

Such benefits actually are significant and include functional and recreational uses, reduced soil erosion and fire danger, and improved infiltration and subsurface water quality. Some of these benefits can be achieved through design alone by using alternate materials. Regulation of turfgrass areas should be based primarily on function. Because turf is widely used and typically over-irrigated, its appropriate use in specific areas should be defined and regulated.

This approach of advocating practical turf areas goes beyond the concept of limited turf areas and includes strategies to reduce turfgrass irrigation. "Practical turf areas promote the use of turf only in those areas of the landscape which provide function" (Welsh 1991). Such functions include recreation and sports; reduced dust, noise, and glare; and temperature mitigation.

Local water-conservation goals and the potential for conservation should determine the level of restrictions. Typically, regulations limit the use of turfgrass from 10 percent to 25 percent of the total landscape area. Such regulations need to include exceptions for parks and recreational facilities to preserve functional amenities.

Another way of limiting turf areas is to use the formula, turf perimeter/area ratio. It was developed to regulate the form turf areas can take. Because narrow shapes are more difficult to irrigate efficiently with spray irrigation, median and parkway strips of turfgrass should be avoided. The turf perimeter/area ratio attempts to preclude long, narrow strips by restricting an allowable relationship between the area and the total perimeter. However, mathematically, long and narrow dimensions can result from using this formula, making it ineffective in restricting all narrow dimensions. The issue of irrigating narrow and odd-shaped areas is better addressed through irrigation standards, discussed later in this chapter.

One other approach is to use water-conserving varieties of turfgrass that have been hybridized to replace more commonly used varieties that consume more water. For instance, in mild-winter climates, tall fescue has been developed that can use as much water but is more drought tolerant than more commonly used bluegrass. Water-conserving varieties are characterized by deeper root systems and improved ability to survive extended periods of reduced irrigation.

Regional patterns of turfgrass use will indicate appropriate standards for water-conserving varieties. This type of regulation offers the benefit of improving water efficiency while preserving the amenities turfgrass provides.

IRRIGATION STANDARDS

Before discussing the types and designs of different irrigation systems, it's important to review some general irrigation standards.

Dedicated Metering or Submetering

Metering of water consumption is essential in implementing a water budget. Where irrigation is not separately metered or a jurisdiction is not metered at all, a checklist approach is necessary. However, separate metering of irrigation use does enhance the checklist approach because management is the most significant and difficult aspect of overall water efficiency to regulate.

While the benefits of metering irrigation use are apparent, many communities are not metered. Converting existing landscapes to metered irrigation is complicated by connection fees and system-capacity charges that must be charged when meters are installed. Strong resistance to mandatory metering can be expected from consumers in nonmetered communities. Metering of new landscapes, however, provides economic incentives and leads to improved management efficiency.

Irrigation Zones

Requiring separate irrigation zones is consistent with grouping similar plants in hydrozoning. An irrigation circuit is a portion of an irrigation system operated by a remote-control valve (RCV). An irrigation zone, or hydrozone, is a portion of the landscaped area having plants with similar irrigation requirements that are served by a valve or a set of valves (Figure 5-1).

Because required irrigation zones make scheduling possible based on actual water requirements for plants, they should be a part of a regulatory approach to standards. While zones do not ensure proper scheduling, they do improve the ability to precisely apply water.

Automating Irrigation Valves

Automating irrigation valves can facilitate efficient irrigation scheduling, especially for large landscapes. Plants requiring irrigation only at monthly intervals should not be automated unless the controller has the capacity to schedule monthly intervals. Where appropriate, RCVs operated by an electronic controller will save labor and allow implementation of predetermined irrigation schedules. Specifying the use of RCVs is common practice in irrigation design because of the obvious benefits.

The use of automatic valves does not imply that a system is fully automatic. Controllers should be reprogrammed regularly, if not linked to real-time weather data, to reflect seasonal changes in water requirements and precipitation. Systems can be automated further by installing automatic rain shutoff devices or moisture sensors for each controller or by computerized central control systems. Central controllers can be linked to weather stations to base schedules on real-time weather data.

Automatic valves should be required to operate all irrigation circuits in new and existing installations if they are likely to facilitate accurate scheduling. However, automation is more likely to reduce irrigation efficiency for residential areas (Wagner 1992). This occurs both because automated irrigation is so easy to use and because of improper scheduling. Irrigation by hose-end sprinklers and watering by hand probably discourages excess application.

Rain shutoff devices. Rain shutoff devices are used to prevent irrigation during periods of rain. The devices are wired to irrigation controllers to override scheduled irrigation. Override occurs when captured rain water opens an electrical circuit. When rain water evaporates from the device, scheduled irrigation resumes. These devices will improve irrigation management because it is not practical for managers of large irrigated areas to manually override irrigation for every controller. Also, they are useful for residential landscapes where irrigation may not be actively managed. However, these devices do not eliminate the need for regular irrigation programming.

Soil moisture sensors. Soil moisture sensors can eliminate much of the need for regular irrigation reprogramming. These devices are placed in the soil at the plant's root zone and wired to an irrigation controller. The controller is programmed with irrigation schedules to meet peak seasonal demand, and the soil moisture level is selected to meet plants' water requirements. Override of scheduled irrigation occurs when the sensor indicates adequate moisture is present.

While soil moisture sensors can automate irrigation scheduling, regular inspection and maintenance is necessary to ensure proper function. In addition, proper

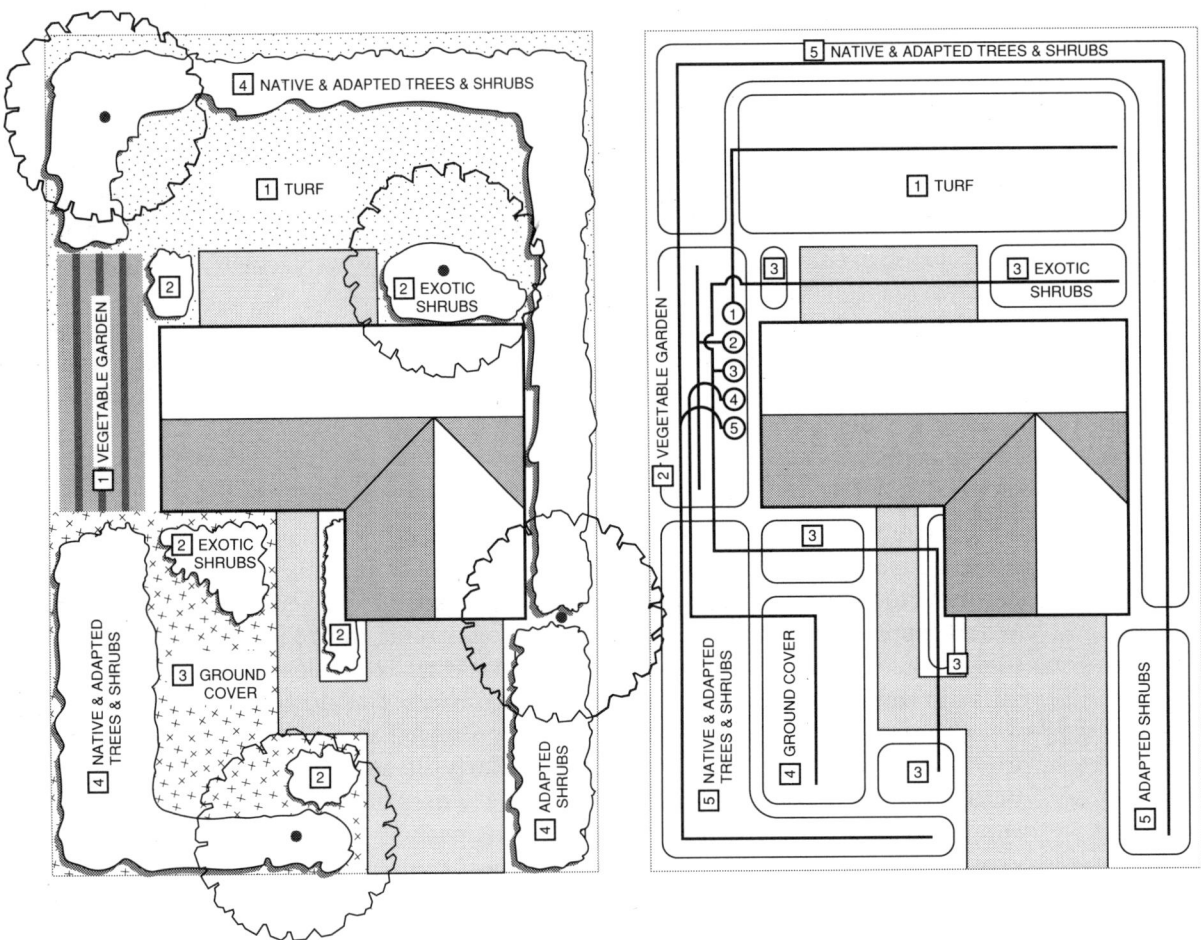

Zone 1	Turfgrass and vegetable garden with high water requirements.
Zone 2	Exotic shrubs with moderate water requirements.
Zone 3	Continuously rooting ground cover with low water requirements.
Zone 4	Native adapted trees and shrubs requiring supplemental irrigation only during the establishment period.

Circuit 1	Spray irrigation for turfgrass.
Circuit 2	Drip system for vegetable garden.
Circuit 3	Drip irrigation for exotic shrubs.
Circuit 4	Drip irrigation with spray-type emitters or subsurface drip irrigation for continuously rooting ground cover.
Circuit 5	Temporary dripline or hand watering during establishment period of native and adapted trees and shrubs.

NOTE: Planning distinct and contiguous plant groupings by similar water requirements allows efficient irrigation with a minimum number of irrigation circuits. Often, more circuits than hydrozones are required to accommodate various plant types and large areas. A large area of turfgrass usually requires more than one irrigation circuit. In the above example, the turf and vegetable garden have similar water requirements but require different irrigation circuits for efficient irrigation.

Adapted from Ferguson (1988).

Figure 5-1 Hydrozones and irrigation circuits for a proposed single-family residence

installation is critical and requires an understanding of water movement in the soil, plant root depths, and plant water requirements. Where sensors are installed, they are often disconnected because the landscape personnel have not been trained to inspect and maintain them.

Many irrigation hardware improvements, such as matched precipitation rate sprinkler heads, are important to hardware efficiency but do not impact scheduling decisions. However, because soil moisture sensors directly impact irrigation scheduling, they have high potential for improving management efficiency.

Irrigation control. Irrigation control should include dual or multiple programming capability and multiple start times to maximize the benefits of automatic valves. Dual/multiple programming is the ability to program irrigation circuits individually. If all circuits are on the same program, it is impossible to accommodate varying water requirements of a landscape. The ability to separately irrigate plant groups requires multiple programs.

Not all irrigation circuits will require separate programs, but each hydrozone should have a separate program since a hydrozone may be served by more than one circuit. A controller should have enough programming capacity to individually schedule all hydrozones.

In addition, multiple start times are necessary to control the duration of irrigation cycles and prevent runoff. Soil infiltration rates often are lower than application rates, so runoff may be generated quickly during irrigation. To reduce wasteful runoff, irrigation cycles can be divided into shorter events using multiple start times. Multiple irrigation cycles can be scheduled with adequate intervals to allow infiltration to the depth of the root zone. This feature is especially important for irrigating slopes and heavy clay soil.

Multiple-program and multiple-start-time controllers should be required on irrigation plans. Because most sites carry a high potential for generating runoff, multiple programming should be mandatory.

Percent switches. Many newer controllers are equipped with a percent switch. This feature controls the duration of irrigation within a program without reprogramming individual circuits. An irrigation schedule for peak demand can be programmed, then easily reduced by an appropriate percentage for periods when demand is lower.

Percent switches simplify reprogramming. A considerable amount of time can be saved where landscape managers are reprogramming many controllers weekly. However, correct duration of irrigation is primarily a function of soil type and the plants' root depth. Frequency rather than duration should be adjusted to accommodate seasonal changes in water requirements.

Correct programming of irrigation frequency is necessary to achieve a professional standard for scheduling efficiency. Exclusive use of percent switches for reprogramming may result in shallow application of water in the root zone. However, the effect of shallow irrigation depth may be offset by higher moisture content remaining in the soil from the previous application during periods of reduced demand.

Despite possible drawbacks to relying on percent reductions of duration to adjust schedules, percent switches provide the landscape manager with a useful tool to improve irrigation scheduling efficiency.

Central irrigation control. Central irrigation control is the computerized control of programming for multiple controllers. Central control uses personal computers and scheduling software to transmit programming information to individual controllers in the field via radio waves or hard wiring. The cost of this technology makes it more appropriate for large, intensely managed sites, such as parks, golf courses, street medians, and new development of city-managed landscapes.

Using central control, an irrigation specialist can schedule large areas from an office, rather than having maintenance personnel adjusting controllers on-site. Scheduling can be linked directly to actual ET data generated by weather stations.

With central control, the potential for water-use efficiency increases because scheduling can be based on real-time ET rather than historical averages. In addition to scheduling improvements, system function can be monitored by installing feedback devices, such as pressure meters and flowmeters. Thus, malfunctions can be detected from a central location and repaired before water losses occur. Another advantage is that multiple programming and start times can be achieved with central control.

Correct operating pressure. Correct operating pressure for all heads and emitters is essential to system efficiency. Spray-irrigation heads should be operated under a range of water pressure specified by the manufacturer. Too little pressure will reduce the range the head can cover and compromise uniform distribution of water. Too much pressure will cause the spray to atomize and be subject to wind drift and overspray. Calculations of pressure loss need to be performed during the design process to ensure proper operating pressure. (This is a standard part of professional irrigation design.)

Pressure reducers are a standard component of most drip systems. Drip-irrigation emitters generally are operated at a lower pressure than sprinkler heads. Predictable flow rates for drip emitters depend on operating them under the water pressure specified by the manufacturer. Drip systems constructed from polyethylene tubing are connected with compression fittings that often fail under too much pressure. Thus, pressure reduction is essential where compression fittings are used.

The certification stamp of a landscape architect or certified irrigation designer on an irrigation plan usually verifies that calculations have been made to assure proper operating pressure. Even so, regulations on correct operating pressure will require irrigation designers to perform a proper analysis. This is vitally important in states where irrigation design is not regulated as a licensed profession.

Reviewing irrigation plans for compliance with this requirement is not practical because of the amount of time and technical information necessary to verify the calculations. Compliance is determined more easily as part of a checklist for field inspection.

Check valves. Check valves should be required where elevation differences cause drainage at low heads (Figure 5-2). When irrigation systems are installed on sloped sites, gravity causes water to drain from the system after each irrigation cycle. Water drains and pools at the lowest head of an irrigation circuit sometimes creating soggy conditions. Water losses can be significant, especially where long runs of pipe or tubing terminate at a low point.

Check valves function to prevent drainage at low heads after irrigation cycles, but they must be installed at all low heads where drainage can occur. Check valves can be installed in irrigation lines or may be factory installed in sprinkler-head assemblies.

In cold-winter climates, check valves will trap water in the irrigation lines. Freezing then can damage the components. In such climates, flush valves must be installed at all low points to ensure that all lines can be drained when freezing temperatures are anticipated.

CHECKLIST APPROACH 63

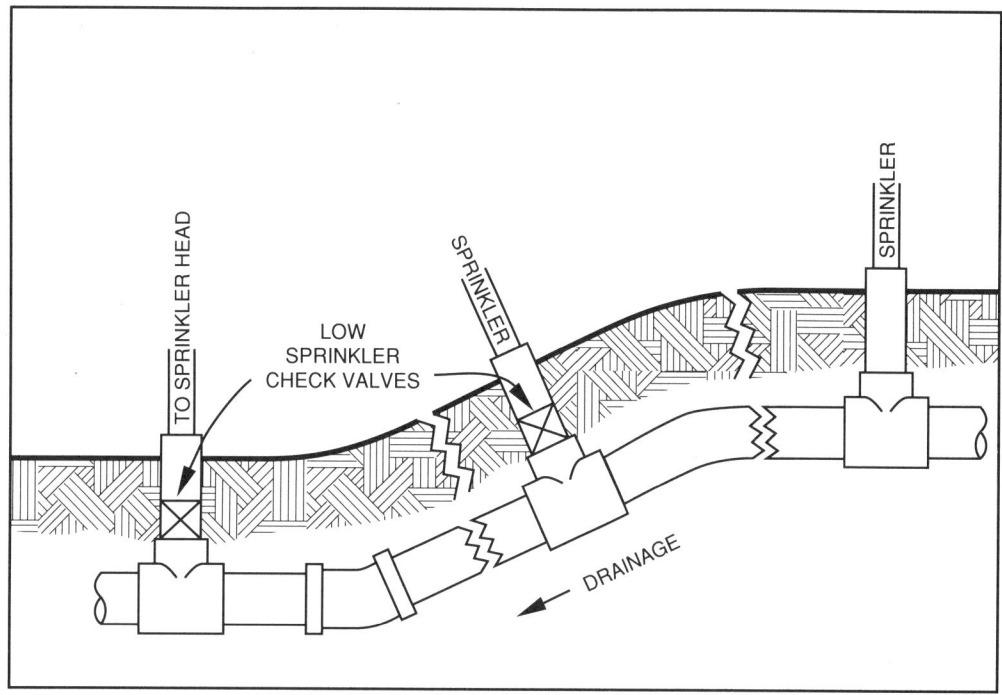

NOTE: Installing a check valve at the lowest elevation will prevent water from draining and pooling at that head and heads down-line at similar elevations. Additional check valves may be necessary upslope to prevent water that is trapped by lower check valves from draining.

Illustration originally published in Turfgrass Water Conservation, *University of California ANR Publications, 1985.*

Figure 5-2 Check valve installation

Irrigation Types

There are two general categories of irrigation: overhead, or sprinkler, irrigation and drip irrigation. Sprinkler irrigation delivers water via bubblers, spray heads, stream rotors, and impact heads. It is characterized by flow rates from the heads measured in gallons per minute (gpm).

Drip irrigation usually delivers water from point-source emitters at the soil surface. However, some drip emitters can be installed below the surface. Other types of drip emitters are similar to fixed-spray irrigation heads because they deliver water from overhead. However, they are considered drip emitters because they deliver water much more slowly and operate with low water pressure. Drip irrigation is characterized by slow, accurate delivery of water measured in gallons per hour (gph).

Drip irrigation is generally not recommended for large areas of turfgrass and continuously rooting ground cover. That is because uniform wetting throughout the entire root zone, required for both these types of plants, is difficult to achieve with drip emitters.

However, new subsurface drip irrigation components have recently become available for this application. It requires careful installation and exact scheduling to be efficient, and improper function is difficult to detect and repair. In addition, scheduling recommendations based on field testing of various emitter spacing, soil types, and plant root depth are not yet available.

Overhead delivery of water is inherently less efficient than drip irrigation due to increased losses from evaporation and overspray. Runoff can be generated quickly on sloped sites. Because of drip irrigation's high potential for efficiency, some ordinances require drip irrigation for all shrubs and trees. This requirement can be complicated because mixed plantings of trees and shrubs with continuously rooting plants are common.

Drip irrigation should be required where sprinkler irrigation efficiency is compromised by site conditions. These conditions include narrow and irregularly shaped areas, shrubbery where foliage interferes with spray patterns, windy sites, and slopes more than 10 percent to 16 percent. Runoff can occur at sites irrigated by drip systems but is easily controlled by constructing watering basins around emitters and installing subsurface drip irrigation systems.

Design Criteria for Sprinkler Irrigation Systems

The following criteria apply specifically to the design of sprinkler irrigation systems.

Head-to-head spacing. Head-to-head spacing refers to the necessary overlap of sprinkler head patterns. The overlap compensates for uneven distribution typical of a single head (Figure 5-3). The precipitation from a head should extend to the next closest head. The precipitation also needs to overlap the next closest head's coverage by 50 percent of the head diameter for square spacing and 55 percent for triangular spacing.

Head-to-head spacing should be required for all irrigated turfgrass and continuously rooting ground cover. It is usually standard in professional irrigation design. Of course, some flexibility is necessary to accommodate irregular site dimensions that do not conveniently correspond to multiples of the radius for standard sprinkler heads. Adequate uniformity can be achieved with spacing up to 1.1 to 1.2 times the head's radius.

Precipitation rate. Precipitation rate, or application rate, is a sprinkler system's delivery rate, expressed in inches per hour (in./h). Flow rate is the volume of water dispensed by a sprinkler head in gallons per minute (gpm).

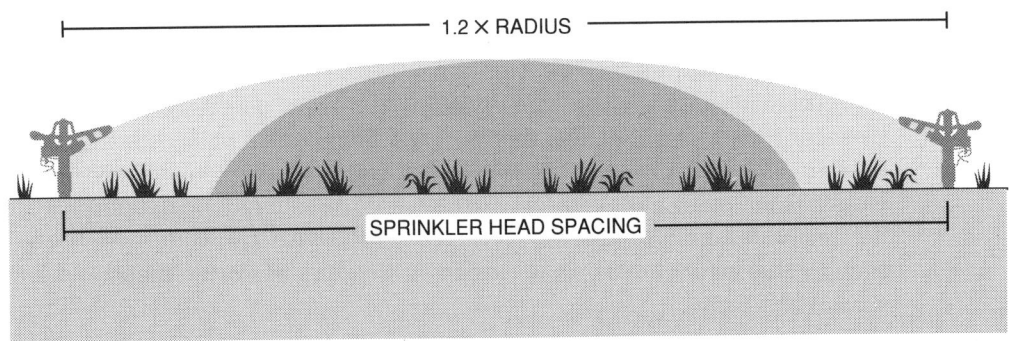

NOTE: The spray of a sprinkler head should extend to adjacent heads. Stretching the spacing up to 1.2× the radius will not seriously compromise uniform distribution of water.

Source: Rain Bird Company.

Figure 5-3 Head-to-head coverage

In order to properly schedule irrigation, a system's precipitation rate must be known. Low precipitation rates, less than 1 in./h, match the low infiltration rates of most soils. The goal is to achieve a precipitation rate lower than the infiltration rate to minimize runoff.

It is not always practical to require precipitation rates lower than infiltration rates. Most soils absorb water very slowly, so preventing runoff is achieved through a combination of minimizing the precipitation rate and scheduling multiple start times.

A lag time occurs between the beginning of irrigation and the generation of runoff. If irrigation cycles can be shortened to prevent runoff, precipitation rates higher than infiltration rates will not reduce efficiency. It is best, though, to design systems with low precipitation rates. Low precipitation rates will improve scheduling options and minimize scheduling errors.

Impact heads and stream rotors have lower precipitation rates than spray heads because they apply water to a much larger area. (Although low-volume spray heads imply lower precipitation rates, low-volume spray heads are fixed spray heads with lower flow rates than standard spray heads. The term "low volume" should not be confused with low precipitation.) Technical specifications usually reveal that low-volume spray heads cover a smaller area than standard spray heads. This results in precipitation rates comparable to standard spray-irrigation heads. Where low precipitation rates will reduce runoff of water from slopes and clay soils, low precipitation, rather than low flow, should be specified.

Matched precipitation rates. Matched precipitation rates enable the designer to select any arc pattern with assurance that all sprinkler heads will combine to produce a uniform precipitation rate (Figure 5-4). Matched precipitation, combined with head-to-head spacing, ensures a professional standard of uniformity and should be required within each irrigation circuit. Existing landscapes can and should be upgraded when a water audit reveals substandard distribution uniformity due to unmatched precipitation rates.

Distribution uniformity. Distribution uniformity is currently tested in the field, rather than in the design stage. New computer programs eventually may make evaluation of distribution uniformity in the design stage more practical. Acceptable uniformity is achieved through proper design, which includes correct operating pressures, head-to-head spacing, and matched precipitation rates. Regular maintenance also is necessary to ensure a high level of uniformity.

Regulation of distribution uniformity in the design stages would require a detailed analysis of the design and the specifications for the components. This analysis could be required as a part of the design and verified by the enforcing agency. A regulation on distribution uniformity should include standards for head-to-head spacing and matched precipitation rates. In addition, a standardized format and procedure for analyzing designs for uniformity would have to be established.

Acceptable levels for distribution uniformity have been established from field testing. Uniformity for turfgrass is more critical to efficiency than for shrub and ground cover areas. That is because turfgrass requires even moisture throughout the planted area to maintain an optimum appearance. But shrubs and trees can draw moisture from a larger root area. Acceptable minimum levels of uniformity are 50 percent for deeply rooted trees and shrubs, 60 percent for shrub and ground cover areas, and 70 percent for turf areas.

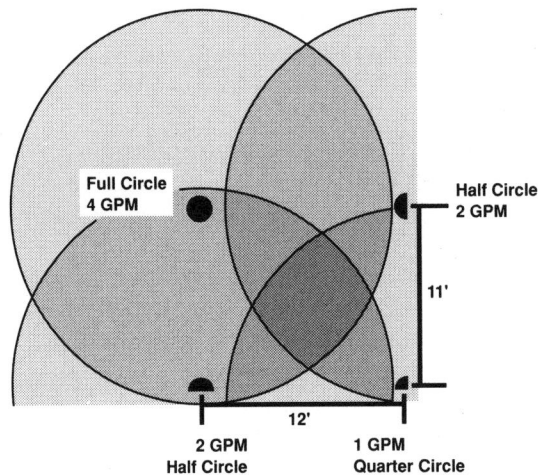

NOTE: Sprinkler heads with matched precipitation rates facilitate uniform delivery of water when varied arc patterns are used to target an area. A half circle will have half the flow rate of a full circle. A quarter circle will have one-quarter the flow rate of a full circle. All heads should be from a single manufacturer's series, since mixing types and brands will compromise uniformity.

Source: Rain Bird Company.

Figure 5-4 Matched precipitation rate sprinklers

Design Criteria for Drip Irrigation Systems

The following efficiency criteria apply to drip irrigation systems.

Filtration. Filtration prevents particles in the water supply from clogging emitters. Screen filters of various mesh sizes are required, depending on type of emitter. Proper filtration and good emitter design have largely overcome the clogging problems once common in drip irrigation. Filtration should meet the standards established by the manufacturer's specifications.

Root zone wetting. Root zone wetting is a function of emitter placement, flow rates, soil types, and irrigation scheduling. A drip system should wet at least 50 percent of the root zone for shrubs and trees. Where possible, all emitters in an irrigation circuit should have the same flow rate. However, because ornamental landscapes are generally mixed plantings, it is advantageous to provide larger trees and shrubs with more emitters and higher flow rates. The appropriate amount of water then can be delivered to plants of various sizes on the same irrigation circuit.

Subsurface installation. Only emitters specifically designed for subsurface installation should be installed below grade; all other emission points should be installed above grade. Also, soil-moisture sensors should be required for subsurface installations. Subsurface systems are more difficult to monitor, so automatic override of irrigation cycles is necessary to prevent undetected malfunction.

Pressure-compensating emitters. Pressure-compensating emitters provide consistent flow rates under a range of operating pressures. This is essential on sloped sites where gravity will increase or decrease operating pressure to individual emitters.

Protection from damage and vandalism. Components of drip irrigation systems in areas subject to foot traffic and vandalism must be protected. Drip systems can be damaged easily because they are typically constructed from polyethylene tubing and compression fittings and installed at the finished grade. A layer of

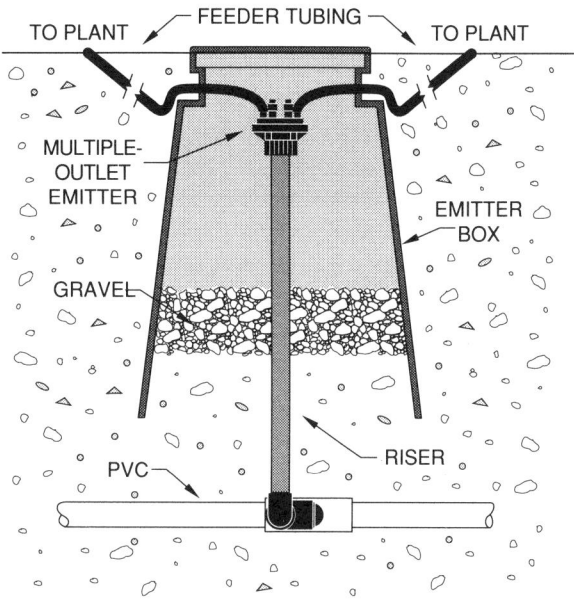

NOTE: Multiple-outlet emitters can be installed below grade in a valve or emitter box to keep them from view and potential damage. Emission points of feeder tubing should be located above grade and the root zones of the plants they serve.

Source: East Bay Municipal Utility District (1991).

Figure 5-5 Multiple-outlet emitter in emitter box with PVC hard pipe

mulch reduces possible damage but does not adequately protect installations in pedestrian areas. Opting for rigid polyvinyl chloride (PVC) pipe and installing protected multiple-outlet emitters can reduce damage (Figure 5-5).

MAINTENANCE STANDARDS

Landscape maintenance can contribute to water conservation in the landscape though it is difficult to quantify actual savings. Timely repair of leaks and breaks prevent water losses that would negate any water savings realized through efficient design and scheduling. Proper installation with quality materials will prevent most problems if systems are maintained regularly.

A checklist approach can specify maintenance procedures and schedules. While landscape plans and specifications can include maintenance schedules, plan review by an enforcing agency does not ensure compliance. An ongoing site-inspection program is required to monitor compliance.

For example, periodically, certified professionals should be required (as part of a regulatory approach) to perform landscape water audits. A water-audit checklist can include the inspection of irrigation systems and landscape maintenance.

The benefits of most maintenance procedures will vary with site conditions and the quality of design and construction. It is more practical to use a checklist to identify serious problems on a site and provide support and materials to landscape managers, than to mandate maintenance procedures that cannot be verified easily.

Turfgrass Maintenance

The following criteria apply to maintenance of turfgrass.

Mowing heights. Mowing heights for turfgrass will affect water requirements (Table 5-2). When turfgrass is allowed to grow to an optimum height, less frequent irrigation may be needed. And, although increasing the mowing height reduces evaporation from the soil surface, grasses that are not mowed will transpire more water due to increased leaf surface area. Optimum mowing heights should be established for the species prominently used in a given region.

Retaining grass clippings. Retaining clippings also will reduce some evaporation from the soil surface. Additional benefits are reduced waste generated by a landscape and the generation of organic compost. It may not be acceptable to leave clippings on smaller, frequently used areas of lawn. Actual water savings from this practice is unclear. Retention of clippings should be left to the discretion of the landscape manager rather than be required by regulation.

Coring and vertical mowing. Coring and vertical mowing are methods of turfgrass cultivation that can contribute to water efficiency by alleviating problems of thatch buildup and surface compaction. Thatch is the accumulation of a layer of organic material at the base of the leaf blades. Thatch on a poorly maintained lawn can become several inches thick. This reduces the vigor of the turf and dramatically reduces infiltration capacity. Surface compaction of the soil, caused by heavy traffic, especially on wet turf, also will reduce infiltration capacity. Low infiltration rates associated with thatch and soil compaction contribute to wasteful runoff during irrigation.

Coring, often called aerification, is the practice by which hollow tines are used to extract cores from the turf. Coring can increase infiltration capacity substantially, especially where surface compaction or thatch limits infiltration. To maximize the benefits of coring, cored material can be removed and holes filled by top dressing with sand or organic compost. For large areas, cores may be left in place and reincorporated into the turf as they break down due to traffic and natural processes (Turgeon 1985).

Vertical mowing is a cultivation procedure involving the use of vertically oriented knives mounted on a rapidly rotating horizontal shaft. Shallow vertical mowing will hasten the breakdown of cored material, which then makes the cored material a more suitable medium for supporting turfgrass growth. Deeper

Table 5-2 Mowing height ranges for turfgrasses commonly used in California

Turfgrass Species	Cutting Height Range, *in.*
Cool-Season Turfgrasses	
Creeping bentgrass	0.2–0.5
Colonial bentgrass	0.5–1.0
Red fescue	1.0–2.0
Kentucky bluegrass	1.5–2.5
Perennial ryegrass	1.5–2.5
Tall fescue	1.5–3.0
Warm-Season Turfgrasses	
Bermudagrass	0.5–1.0
Zoysiagrass	0.5–1.0
Seashore paspalum	0.5–1.0
St. Augustine grass	0.5–1.5
Kikuyugrass	0.5–1.0

Source: Gibeault (1991).

penetration will remove accumulated thatch. And, deep vertical mowing, where knife blades penetrate underlying soil, may alleviate surface compaction (Turgeon 1985).

Coring and vertical mowing work for selected sites. Sloped sites will require more maintenance to prevent runoff. The need for these procedures should be identified as part of a water-audit checklist.

Irrigation System Maintenance

System maintenance is essential to the efficiency of automated irrigation systems. "Even the most sophisticated irrigation equipment is rendered useless unless there is a firm commitment to proper system maintenance" (American Society of Irrigation Consultants).

Automation, however, reduces maintenance personnel's observation of irrigation systems during operating hours. Specification of system maintenance, as with other maintenance procedures, does not ensure compliance, and inspection of the site will reveal problems only after they have occurred.

The most basic maintenance procedure is a regularly scheduled, daytime, visual inspection of a system while it is operating. This procedure usually is not scheduled or itemized in landscape maintenance estimates. However, landscape specifications should include a maintenance schedule for inspection, testing, and maintenance procedures.

Procedures should include: cleaning of filters and strainers, flushing of irrigation lines, adjusting of sprinkler patterns to maintain uniformity, and calibrating all sensing and recording equipment. System parts should be replaced and upgraded using the same or compatible irrigation components. For example, different brands, manufacturers' series, or types of sprinkler heads should not be used in the same irrigation circuit.

Horticultural Practices

Certain horticultural practices also will increase the water efficiency of a landscape. Because these practices indirectly impact water consumption, the resulting water savings are difficult to quantify. And because the practices will vary with each site, their benefits also will vary.

Mulching. Mulching reduces evaporation from the soil surface and temperature fluctuation in plant root zones. Organic and inorganic materials can be used as mulch, but impervious material such as plastic should not be used. Organic mulch is preferable to inorganic mulch because it benefits plant growth. Organic mulches will moderate soil temperatures and gradually break down and improve soil structure and the availability of nutrients. Both types of mulch also reduce runoff and the frequency of irrigation. A minimum of 2 in. of mulch should be added to the soil surface after planting in nonturf areas.

Fertilizer application. When excessive or inappropriate, fertilizer will increase plants' water requirements by encouraging new and rapid growth. Conversely, over-irrigation will increase the need for fertilizers by leaching nutrients from plant root zones. Excessive applications of fertilizer on turfgrass also can contribute to thatch buildup and reduce infiltration rates.

Fertilizer use may be minimized or discontinued altogether for some established landscapes. Plants naturally adapted to a climate may not require additional fertilizer. As with water application, fertilizer applications should be based on each site's specific, seasonal needs. Fertilizer applications also should be coordinated with irrigation schedules to maximize effectiveness.

Pruning. Pruning and thinning of foliage can increase plants' water requirements, especially in hot climates where shade reduces evaporation. Shading of the ground plane and understory plantings effectively creates microclimates with lower ET rates.

Pest control. Pest control, including weed, insect, and disease control, reduces plants' water requirements and indirectly impacts water efficiency. Proper design and scheduling of irrigation systems facilitates pest control. Weeds and waterborne pests prevail when excess water is applied. Weeds and pests compete with plants for available light, water, and nutrients. They also reduce drought tolerance. However, because of the indirect connection to water efficiency, pest control is beyond the scope of water-efficiency regulation.

SCHEDULING STANDARDS

Proper irrigation scheduling is essential in saving water. For most existing landscapes, improved scheduling will generate more water savings than any other factor of landscape water efficiency. Plan review and site inspection ensure potential for water-use efficiency but do not monitor adherence to irrigation schedules. Detailed irrigation schedules, however, can be generated in a project's design stage and should be required in a regulatory approach.

Schedules should be based on established values for plants' water requirements and system performance. Schedules can be calculated using ET data; standardized data on plants' water requirements; and information from irrigation, planting, and grading plans.

Schedule requirements should be accompanied with work sheets that standardize the process and encompass all related factors. For new installations, monthly schedules should be prepared for the plants' establishment period and for the mature landscape.

A landscape usually requires two or three years to mature. Newly planted shrubs and trees will require shorter, more frequent irrigation than a mature landscape. As a landscape matures, longer, less frequent irrigation will be required. Shallow-rooted plantings, including many varieties of turfgrass, will have shorter establishment periods.

Some landscape ordinances will allow more water consumption during the establishment period. This provision is based on the assumption that a mature landscape requires less water than a new landscape. Depending on the type of plants being irrigated, a new landscape initially may require less water due to less foliage mass and transpiration. Conversely, water-conserving plants may require irrigation during the establishment period and no supplemental water once they are established. Correct irrigation schedules will obviate the need for additional water during the establishment period.

Scheduling Factors

To effectively program an irrigation controller, at least two values must be calculated for each irrigation circuit: the frequency of irrigation and the duration of the irrigation cycle. In varying and sometimes confusing ways, automatic controllers incorporate these values. Frequency is programmed by setting the days of the week to water (days on) or setting the interval of days at which watering will occur (interval). Duration is programmed by setting the minutes for which an irrigation circuit will operate. In addition, the current calendar day and time and the time(s) that irrigation cycles will begin (start times) must be programmed.

All the program information must be coordinated with the days and hours that irrigation can occur; this is sometimes called a watering window. For example, a baseball field can only be irrigated while not in use, and drying time must be allowed before play begins. A watering window is the day and time available for irrigation. Generally, nighttime and early morning hours are preferable for irrigation because less wind and evaporation occurs than during daytime hours.

Irrigation formulas are used to calculate duration, frequency, and the number of daily cycles. (See appendix B for formulas.) Duration is a function of soil-moisture-holding capacity, application rate, system efficiency, soil type, and plant root depth. Frequency is derived from the soil-moisture-holding capacity and the rate at which the soil dries out due to evapotranspiration. The number of daily cycles is a function of the soil's infiltration rate and slope. Established formulas found in currently available irrigation scheduling manuals will serve to generate adequate schedules (Toro Company 1986; Rain Bird Company; Shepersky 1984).

Additional factors to consider are the mix of plant species, density of plantings, and microclimates. The variety of factors and multiple formulas can complicate the scheduling process, but a workable methodology that incorporates these additional factors is available for calculating precise estimates of water use and irrigation schedules (MacNair 1991).

The landscape coefficient method, developed by the University of California Cooperative Extension, establishes a plant factor for an irrigated area. This factor is similar to the plant factor K_c described in chapter 4 but also accounts for the density, microclimate, and the plant species. Once the ET is known for a given area, frequency and duration of irrigation cycles can be determined for irrigation circuits.

Irrigation scheduling software. The complexity of irrigation scheduling for ornamental landscapes results from the diversity and mix of plant species, microclimatic influences, planting densities, and the varied types of irrigation equipment and site conditions.

Although precise monthly schedules can be generated that account for all these factors, this type of detailed irrigation scheduling is time-consuming and is not a standard part of irrigation design. Advances in irrigation scheduling include computer software that can help generate schedules. This technology can contribute to the standardization of scheduling methodology as required by regulation.

The value of precise scheduling methodology will be realized only if the schedules are implemented in the field. Design and scheduling information needs to be transferred to the landscape managers. This can be problematic because separate contractors usually handle design, installation, and management of a site. One solution would be to require by regulation that irrigation schedules be submitted to the reviewing agency and this information be transferred to the landscape managers.

REFERENCES

American Society of Irrigation Consultants. *Minimum Standards For Landscape Irrigation.* ASIC, Lafayette, Calif.

Costello, L.R. & Jones, K.S. 1992. WUCOLS Project: Water Use Classification Project. University of California Cooperative Extension, San Francisco and San Mateo County Office, Half Moon Bay, Calif.

East Bay Municipal Utility District. 1991. *Drip Irrigation Guidelines.* Office of Water Conservation, Oakland, Calif.

Ferguson, B.K. 1988. "Using Water Effectively." In *Irrigation, Volume 3 of Handbook of Landscape Architectural Construction.* Edited by Scott Weinberg and John Mack Roberts. Landscape Architecture Foundation, Washington, D.C.

Ferguson, B.K. & Debo, T.N. 1990. *On-Site Stormwater Management*. Van Nostrand Reinhold, New York.

Gibeault, V.A. et al. 1991. Managing Turfgrasses During Drought. University of California Cooperative Extention Leaflet 21499, Division of Agriculture and Natural Resources, Oakland, Calif. (June 1991).

Gibeault, V.A. & Cockerham, S.T., eds. 1985. Turfgrass Water Conservation. University of California Cooperative Extentions, Division of Agriculture and Natural Resources, Oakland, Calif.

MacNair, James. 1991. Irrigation of Nonturfgrass Landscape Areas. Irrigation Management Group, Union City, Calif. (November 1991).

Rain Bird Company. Landscape Irrigation Design Manual. Rain Bird Sales, Inc., Turf Division, Glendora, Calif.

Shepersky, Keith. 1984. The Rain Bird Landscape Drip Irrigation Design Manual. Rain Bird Sales, Inc., Turf Division, Glendora, Calif.

Toro Company. 1986. Automatic Sprinkler System Scheduling: The Educator Series Vol: 1, No. 1. The Toro Company Irrigation Division, Toro Company, Riverside, Calif.

Turgeon, A.J. 1985. *Turfgrass Management*. Reston Publishing Co., Reston, Va.

Wagner, Christina. 1992. Residential Landscape Comparison Study. East Bay Municipal Utility District Water Conservation Office, Oakland, Calif. (August 1992).

Welsh, D.F. 1991. Practical Turf Areas: The Controversial Xeriscape Guideline. *Turf News*, Special Issue. pp. 47–48, 50, 58, 60. (September 1991).

6

Implementing Landscape Standards

Water budget and checklist approaches to standards can be implemented as guidelines through either educational or regulatory programs. Implementing regulations based on a water budget approach requires the involvement of the local water agency (city water departments, water districts, and private water purveyors). Implementing regulations based on a checklist approach requires the involvement of a water utility or a city or county agency.

A comprehensive approach to implementation includes:

- informing landscape industry professionals, agency staff, and consumers of the requirements for compliance and the consequences of noncompliance
- facilitating compliance and the exchange of information
- enforcing the consequences to ensure compliance

Local regulatory and administrative structures are usually in place through city and county planning and building departments. Regulatory programs can be tied to existing permit and inspection procedures. Water-budget programs can be implemented by water agencies through rate structures and the regulation of water service. Either approach requires a strong educational program to bridge the gap between the potential for water-use efficiency and actual, current landscape practice.

The responsibility for regulation lies in the public sector, but its implementation should involve the private sector in two ways: through advisory committees that work with government agencies to develop standards and through educational programs and services offered by local agencies.

A cooperative effort to address the increasing demands on water supplies benefits both the public and private sectors. The public sector benefits through improved regulation and compliance. The private sector benefits through higher standards for professional practice and improved cost-effectiveness.

IMPLEMENTING THE WATER BUDGET APPROACH

By definition, implementing a water budget requires access to consumption data, which means involving a water utility. Depending on the administrative structure, planning agencies and other government agencies may acquire the data needed to assess compliance.

Submittal and Review

A major advantage of the water budget approach is the simplification of the submittal and review process for landscape plans. Only a minimum amount of information needs to be exchanged to establish a water allowance for a given site, but this information should be documented and verifiable.

For example, to complete the submittal and review process, irrigated area (IA) or landscaped area (LA) must be known, as it is the defining variable in Eq 4-2, the water budget formula. The LA and IA will vary with each site. See chapter 4 for more detailed information.

Landscape and irrigated areas are measured from plans drawn to scale, aerial photos, or site surveys. IAs should be calculated for each meter. To determine IA per meter from a plan, the irrigation valves can be traced back to their corresponding meter. The area served by each meter then can be measured. Measurements are done with architects' or engineers' scales or a planimeter can be used for irregular areas.

Aerial photos can be used for larger sites where plans are not available. They can be purchased from a city's engineering or planning department or from a local aerial photo company. Area takeoff from photos is calculated as it is with plans drawn to scale. Area also can be measured in the field by qualified personnel.

Documentation. Documentation of irrigated acreage also should be required in the form of as-built irrigation plans that are drawn to scale. These as-built plans should be construction drawings that reflect any changes made to the design during installation. Plans also should include meter location(s) and the delineation of area(s) served by a meter(s).

Aerial photos and site surveys can substitute for plans, but they lack the delineation of subsurface irrigation components. Also, in the absence of irrigation plans, it may be difficult to determine the area served by each meter. IA for each meter can be determined by field observation of system and irrigation meter operation. If a field survey is used to measure area, a diagrammatic plan documenting the survey should be prepared to substantiate calculations.

To facilitate the exchange of information, forms should be prepared that clarify the required submittal and documentation. The documentation then should be reviewed to ensure accuracy of the submitted square footage. Reproducing area calculations can be very time-consuming and some margin of error will exist for all calculations. However, a quick check can be performed to ascertain the accuracy of a submittal. For example, 10 percent of the sites in a submittal can be randomly checked to verify correctness of the documentation and provide an incentive for accurate submittal.

Calculations. Water budget calculations can be performed using Eq 4-1. Using Eq 4-2, an adjustment factor also can be included if needed to reflect local conditions and water-conservation goals.

An annual budget needs to be distributed throughout the year to reflect seasonal irrigation needs. To distribute a budget throughout existing billing periods, monthly ET figures can be used as shown in Eq 4-3. Computer spreadsheets will facilitate this process. In fact, computer programming may be necessary to integrate the water-budget figures into the water purveyor's billing system.

Water Banking and Pooling

Other, more flexible mechanisms can also help in distributing water budgets over 12 months. Water banking and pooling of water allotments are two such tools that water utilities can use to provide flexibility while preserving an overall incentive to conserve. Two disadvantages are that these mechanisms do require additional administrative time and need to be compatible with the existing billing system.

Water banking. Water banking allows unused water allotments to be credited to another billing period within the same year. Water banking may be allowed for each billing period or, to reduce paper work, only three to four times a year. Shifting water allotments from one billing period to another better accommodates natural variations from the average, historic ET data. Water banking also can strengthen the incentive for landscape managers to irrigate efficiently.

Pooling. Pooling is a process whereby allocations from several different irrigation accounts are reassigned to better fit actual watering needs. The total amount of allocated water does not increase; it is merely redistributed amongst multiple accounts. This better accommodates landscape renovation and changes in use that increase the need for supplemental irrigation. It also allows the landscape manager to establish priorities for irrigation use. For larger sites with multiple accounts, fringe areas can be placed under reduced irrigation, while areas of intense use and plants that need a lot of water can receive additional water.

A water budget approach is not usually consistent with pooling because water budgets are site-specific. One reason for this is that if water budgets for all sites are adequate for efficiently irrigating high-water-use plants, additional water for any one site is unnecessary. And, additional water for a given site, obtained by pooling the water budget with another low-water-use site, dilutes the incentive to irrigate efficiently.

Where strict efficiency standards imply limits on high-water-use plants, pooling does not eliminate the incentive to increase efficiency. Pooling may be used to prioritize water use and preserve the recreational and functional amenities of large turf areas. For example, a city may choose to install water-conserving plants that require minimal irrigation in street medians and then reassign some of the water allotment to sports fields that have high water requirements.

Looped pooling. Looped pooling allows the total allotments and total consumption for several accounts to be compared. If total use exceeds the total allotment, penalties take effect. Looped pooling helps accommodate irrigators with more than one meter serving the same irrigation system. (Irrigation meters sometimes are looped together to serve the same irrigation main.) Depending on the system's pressure demands, more or less water will be drawn through a meter. So, calculating the irrigated area served by a single meter in a looped system is problematic. All accounts serving a loop should be handled as one account for water-budgeting purposes.

Factored pooling. Factored pooling is a process of reassigning allocations using an adjustment factor to calculate allotments before consumption and billing. This method is advantageous if billing adjustments based on actual consumption are impractical because of the billing system and administrative capacity.

Factored pooling also requires customers to monitor consumption and to project a rate of use for each account. Many landscape managers don't know what consumption patterns are, so this type of pooling fills an important educational need by requiring managers to research water needs and project rates of use. Factored pooling calculations are explained in appendix B.

Enforcement Mechanisms

The consequences for exceeding established water budgets can take the form of water-service regulations and rate structures.

Water-service regulations. Water-service regulations are contingencies for water service. Water service can be denied, terminated, or limited by flow restrictors for noncompliance to a water budget. These are very severe options for a site that depends on irrigation. They could even cause loss of landscaping. Flow restrictors will render most irrigation systems nonfunctional and could have the same effect as termination of service.

These kinds of contingencies are more suited to the checklist approach and for new landscapes where service connections are just being established. Landscape construction requires water service, so contingencies would have to be established early in the planning stages. If a water budget must be submitted for new service, a special allowance for construction use would have to be provided. Any changes made in total irrigated area during installation would require an adjustment to the water budget.

Water-conserving rate structures. Water-conserving rate structures are the primary mechanism of enforcement in the water budget approach. Enforcement through rate structures assumes that dedicated metering exists for all irrigation connections and billing is done by volume of use. Penalties for excessive water use provide the incentive for efficiency.

Water is usually the most inexpensive utility for residential customers, so small price increases for these types of customers will not significantly impact the total cost of utilities. Steep inclining rate structures (rate structures on a sliding scale) that penalize excessive water use will have the most impact on residential customers with extensive landscapes. Increased rates for large irrigators, however, will have a significant impact on landscape maintenance costs and are an effective means to achieve water-conservation goals.

IMPLEMENTING THE CHECKLIST APPROACH

Under the checklist approach, standards can be implemented by public agencies, including water utilities, city planning departments, and building inspection agencies. The enforcing entity must have the legal authority (and not just the responsibility) to implement regulations. City planning departments can link compliance to existing permit and inspection processes. And, water utilities can implement water-service contingencies through their new-service and water-conservation offices.

Where feasible, it may be advantageous for regional water utilities to enforce regulatory programs. Requests for new water service provide an opportunity to regulate projects in the preplanning and planning stages. However, local government entities may be reluctant to yield regulatory authority to a regional water district. Furthermore, the water agency may not have the necessary administrative capacity.

The checklist approach requires both detailed analysis of plans and site inspections for compliance. Public entities may not have the staffing levels and expertise required for this kind of review and inspection. While increased bureaucracy is not desirable, cost-effective, long-term water conservation may justify increased staffing levels.

Private Sector Signatory

An alternative to increased staffing for public entities is to use licensed, certified landscape professionals in the review and inspection process. Licensed landscape architects and irrigation designers can be required to verify compliance.

Participation of the private sector could be viewed as a boon to landscape professionals because of the additional work it generates and because higher standards can increase professionalism. Some professionals, however, perceive a conflict to their client relationship by being required to police regulation compliance.

In addition to potential conflicts of interest, certain legal issues arise for landscape professionals required to certify compliance. The following areas need to be addressed if landscape professionals and, especially, licensed landscape architects act as signatory to the review and approval process:

- Errors and omissions insurance may not permit licensed landscape architects to certify the correctness of planting and irrigation installations based on field observations.

- In some states, landscape architectural licensing law does not allow licensed architects to certify the work of nonlicensed practitioners. Often, other design specialists are part of a design team along with landscape architects and are not directly supervised.

- Landscape professionals may be exposed to increased liability if regulations are not complied with and loss of the landscape occurs or penalties result.

- Responsibility for compliance by owners, landscape managers, and landscape contractors may not be clearly separated from the design and certification process.

To avoid legal and perceptual problems, landscape architects, rather than certifying compliance, can be required to sign a statement that a substantial amount of the work has been completed. Certification of compliance should be the enforcing agency's purview and should be carried out by inspecting the site for a fee.

Submittal Requirements

Compliance with regulations is accomplished primarily through reviewing landscape plans and inspecting each site. While submittal requirements should adequately document compliance to specific design and scheduling standards, the required document submittal should be minimized in order to facilitate compliance. Standardized forms (provided by the regulatory agency) should be used to facilitate the exchange of information between the regulatory agency and the water user about regulatory requirements and the submitted documents.

Submittal requirements may include all or some of the following components, depending on the nature of the regulations:

- A design concept statement is a general requirement. It explains the function of water-consumptive landscape elements and the use of other elements that conserve water.

- Irrigation, grading, and planting plans may be combined for smaller projects. They are necessary to document compliance with many specific criteria.

- Monthly irrigation schedules outline each circuit's cycles per day, duration and frequency of cycles, precipitation or application rates, and water requirements for plants. If the design is to be analyzed for compliance to a water budget, the schedule also should note each circuit's irrigated area and the plant factor used. Schedules should be prepared for both the establishment period and the mature landscape.
- Total irrigated acreage should be determined as part of the design process and submitted if a landscape design is analyzed for compliance to a water budget. Where percent limits on turfgrass or high-water-use plants are in effect, total square footage of the landscaped area and the square footage of restricted plants should be submitted.
- An annual water budget should be calculated using a formula designated by the regulations.
- A soils report should include infiltration rate(s) and recommendations for soil improvement.
- Schedules of postconstruction review and water audits also can be included.

Enforcement Mechanisms

Several mechanisms can be used to enforce a checklist approach. They are discussed below.

Design review. Design review and approval is the primary mechanism used to ensure compliance to the checklist approach. However, design review does not address compliance of a site's installation or management. A comprehensive program will include inspection of the installation and periodic water audits of system efficiency, scheduling, and maintenance.

Checklists should be developed as part of a systematic, consistent approach to design review. Checklists should address compliance to all the specific standards.

Checklists need to include plant selection. Selection should be verified using locally developed lists of plants and their approximate water requirements. Because empirical data from field measurements is not available for most ornamental plants, lists should be compiled by horticultural consultants using available literature and practical experience.

Including consultants from the landscape industry in the process will facilitate acceptance of a plant list. Lists should be amended and changed based on research findings and new horticultural introductions.

Calculating compliance through water budgets. A checklist approach also can incorporate a water budget to measure compliance. Compliance of the design rather than compliance to consumption limits differentiates the checklist approach from the water budget approach. Water demand generated by the design can be subject to usage limits rather than specific plant types and design features. This provides greater flexibility in design and can reduce the number of specific requirements.

To calculate design compliance to a water budget, water requirements for ornamental plants must be established locally. The lists then are used to estimate a landscape's actual water demand during the planning stages. An analysis of planting, irrigation, and grading plans for each hydrozone is necessary to verify compliance to a water budget.

Equation 4-2 establishes the limits for the entire site. The adjustment factor (AF) establishes the local standard of efficiency for irrigation and for plants' water requirements.

To calculate compliance, the plant factor, irrigation efficiency, and irrigated area must be determined for each hydrozone, then applied to Eq 4-2. The gallons or units for each hydrozone then are totaled and compared to the annual budget.

For example, assuming a local jurisdiction has an annual ET_O of 48 in., a 0.6 adjustment factor, and the landscaped area of the site to be reviewed is 50,000 ft^2, the annual budget calculations are:

$$(48 \text{ in.}) (0.6) (50,000 \text{ ft}^2) (0.623 \text{ gal/ft}^2 \text{ in.}) = 897,120 \text{ gal}$$

The matrix shown in Figure 6-1 presents an example of calculating compliance for a hypothetical landscape of 50,000 ft^2 with four hydrozones.

Hydrozone Description	K_l	IE	AF	IA$_{zone}$ ft^2	ET_O, in.	CF	Annual Gallons
Hydrozone #1 Turfgrass with sprinkler	0.8	0.65	1.23	10,000	48	0.623	367,819
Hydrozone #2 Groundcover with sprinkler	0.6	0.65	0.92	15,000	48	0.623	412,675
Hydrozone #3 Trees, shrubs with drip	0.3	0.9	0.33	5,000	48	0.623	49,342
Hydrozone #4 Nonirrigated	0	NA	NA	20,000	48	0.623	0
Totals				50,000			829,836

NOTES: NA = not applicable.

K_l is the plant factor for the hydrozone. It is derived from the highest plant factor for all the plants in the zone. It should be predetermined and taken from a plant list established for the purpose of plan review. Grouping plants with similar requirements in the same zone will minimize K_l for the entire zone and reduce overall demand.

IE is the irrigation efficiency of the zone. Depending on the type of irrigation used in a hydrozone, different values may be assigned for IE.

AF is the adjustment factor for each zone. K_l/IE = AF

IA$_{zone}$ is the irrigated area for each zone and is calculated by area takeoff from landscape plans.

ET_O is the annual, local, reference evapotranspiration rate in inches.

CF is the conversion factor to convert ET in inches to gallons.

Annual gallons for each zone are calculated by applying the formula to each zone:

$$(ET_O) (AF) (IA_{zone}) (CF) = \text{annual gallons}$$

Figure 6-1 Design compliance calculations

In the preceding example, compliance is achieved because the matrix yields less total annual gallons (829,836 gal) than allowed by the annual water-budget formula applied to the entire site (897,120 gal).

Where water budgets are used to measure compliance to a checklist approach, the above calculations should be performed as part of the design process and submitted for verification. Water budget calculations can be a design tool used to achieve greater efficiency.

Simple computer spreadsheets can be used to facilitate the process. Computer software should be made available to designers and enforcing agencies wherever possible to standardize the process. Work sheets with instructions also can be used and may be more practical for smaller projects or as a substitute for computer technology.

Currently, advances are being made in the development of software for this purpose. Data bases and formulas associated with computer spreadsheets can standardize the assumptions on which calculations are made. This technology has the potential to facilitate compliance as well as educate users to water-efficient design.

Site inspection. Site inspection is necessary to ensure compliance to design requirements. Design changes frequently occur during construction and need to be reviewed for compliance. Depending on the specific requirements, two site inspections may be necessary. An initial inspection may be needed before the irrigation components are covered with backfill. And final inspection should be performed once installation is complete. As with plan review, checklists should be developed to reflect all specific requirements.

Using landscape professionals as signatory to certify that an installation has been done correctly also may be considered. This is discussed above.

Water audits. Water audits should be performed at specified intervals to review system efficiency and maintenance practices. For purposes of discussion, any method of improving landscape efficiency through on-site consulting services is considered a landscape water audit.

Water-audit procedures and training have been developed by some state water agencies. (For more information, see appendix D for a partial list of state-level agencies and contacts.) In addition, private consultants have developed techniques for improving efficiency through meter-reading programs and site surveys.

Water audits, an important educational tool, require the presence of the landscape manager. The time to schedule water audits is at the time of approval after design review. They should be slated for intervals of two to five years. This followup procedure is necessary to address important management issues that are not verifiable in the design review process.

A water auditor will evaluate system efficiency through field measurements and make scheduling recommendations. Because landscapes mature and often change from the original design, upgrades and changes to system hardware and scheduling often are necessary to maintain overall management efficiency.

EDUCATIONAL AND INFORMATIONAL SERVICES

Educational and informational services are also important to the success of regulatory approaches to standards. Extending such services to the landscape industry will not only fill an informational gap but will foster cooperation between the private and public sector.

Examples of an educational campaign include:

- Workshops explaining new regulation and requirements for compliance. Such workshops could cover landscape management and irrigation scheduling.

- The promotion of water audits and system upgrade services. This can be achieved by offering financial incentives to improve irrigation efficiency.

- Informational materials that address the specific needs of a region. Research can be sponsored or conducted to provide an empirical basis for landscape standards.
- Telephone lines with daily ET information from weather stations can provide irrigators with necessary scheduling data. For example, the California Irrigation Management Information System (CIMIS) provides information for agricultural and landscape managers throughout California.

Appendix A

Selected Evapotranspiration and Rainfall Data*

In regions where rainfall exceeds potential evapotranspiration (ET) during most months of the year, supplemental irrigation may be required only during periods of below-average rainfall. In arid western states, supplemental irrigation may be required for long growing seasons characterized by minimal rainfall. Using ET rates to establish a landscape's water budget is more practical where rainfall is consistently below ET during the growing season. And in regions where episodic drought temporarily creates demand, a checklist approach may be more appropriate.

ET will vary from region to region and site to site due to microclimatic conditions. To determine irrigation schedules, reference ET needs to be adjusted for known plant water requirements and site conditions.

The figures represent inches of moisture. Rainfall (RF*) is based on 30-year averages from 1930 to 1961. ET figures are potential evapotranspiration calculated from a modified Blaney–Criddle formula. The difference between RF and ET is designated as DIFF. Data for additional locations can be obtained from the Toro Company or from state agricultural services.

Key

ET   = inches of potential evapotranspiration
RF   = inches of rainfall
DIFF = RF − ET

---

*Information in this appendix was taken from Toro Company, *Rainfall-Evapotranspiration Data, United States and Canada*. (Minneapolis: Toro Company, 1966).

Selected Evapotranspiration and Rainfall Data

ALABAMA		JAN	FEB	MAR	APR	MAY	JUN	JUL	AUG	SEP	OCT	NOV	DEC	TOTAL
PRAIRIE (MONTGOMERY)														
	RF	4.32	4.78	5.88	5.02	3.68	3.76	5.47	4.04	3.23	2.15	3.16	4.85	50.34
	ET	0.95	1.21	2.36	3.93	6.01	7.54	8.10	7.46	5.67	3.59	1.74	1.06	49.62
	DIFF	3.37	3.57	3.52	1.09	-2.33	-3.78	-2.63	-3.42	-2.44	-1.44	1.42	3.79	0.72
GULF (MOBILE)														
	RF	4.44	4.20	6.74	5.95	5.01	5.87	8.93	6.26	6.43	2.91	3.44	5.02	65.20
	ET	1.12	1.42	2.66	4.19	6.16	7.48	7.86	7.56	5.83	3.91	2.00	1.31	51.50
	DIFF	3.32	2.78	4.08	1.76	-1.15	-1.61	1.07	-1.30	0.60	-1.00	1.44	3.71	13.70

ARIZONA		JAN	FEB	MAR	APR	MAY	JUN	JUL	AUG	SEP	OCT	NOV	DEC	TOTAL
NORTHEAST (FLAGSTAFF)														
	RF	1.21	1.25	1.12	0.82	0.50	0.47	1.78	2.34	1.36	1.14	0.78	1.18	13.95
	ET	0.25	0.43	1.05	2.06	3.50	5.20	6.20	5.52	3.81	2.02	0.75	0.35	31.14
	DIFF	0.96	0.82	0.07	-1.24	-3.00	-4.73	-4.42	-3.18	-2.45	-0.88	0.03	0.83	-17.19
SOUTH CENTRAL (PHOENIX)														
	RF	0.97	0.96	0.77	0.39	0.14	0.14	1.06	1.56	0.83	0.59	0.60	1.01	9.02
	ET	0.94	1.31	2.54	4.22	6.42	8.33	10.11	8.97	6.75	4.11	1.89	1.16	56.75
	DIFF	0.03	-0.35	-1.77	-3.83	-6.28	-8.19	-9.05	-7.41	-5.92	-3.52	-1.29	-0.15	-47.73
SOUTHEAST (TUCSON)														
	RF	0.98	1.02	0.73	0.36	0.16	0.49	2.78	3.14	1.34	0.79	0.64	0.99	13.42
	ET	0.76	1.05	2.06	3.50	5.29	7.16	8.11	7.19	5.51	3.47	1.60	0.96	46.66
	DIFF	0.22	-0.03	-1.33	-3.14	-5.13	-6.67	-5.33	-4.05	-4.17	-2.68	-0.96	0.03	-33.24

ARKANSAS		JAN	FEB	MAR	APR	MAY	JUN	JUL	AUG	SEP	OCT	NOV	DEC	TOTAL
CENTRAL (LITTLE ROCK)														
	RF	4.94	4.45	5.01	5.45	5.77	4.04	3.72	3.18	3.33	3.24	4.27	4.26	51.66
	ET	0.60	0.83	1.86	3.52	5.46	7.21	8.19	7.49	5.22	3.20	1.36	0.74	45.68
	DIFF	4.34	3.62	3.15	1.93	0.31	-3.17	-4.47	-4.31	-1.89	0.04	2.91	3.52	5.98

Selected Evapotranspiration and Rainfall Data (continued)

CALIFORNIA		JAN	FEB	MAR	APR	MAY	JUN	JUL	AUG	SEP	OCT	NOV	DEC	TOTAL
NORTH COAST DRAINAGE (EUREKA)														
	RF	7.73	6.35	4.89	2.60	1.68	0.72	0.15	0.15	0.53	2.67	4.42	7.43	39.32
	ET	0.59	0.85	1.60	2.44	3.59	4.51	5.32	4.75	3.68	2.38	1.18	0.71	31.60
	DIFF	7.14	5.50	3.29	0.16	-1.91	-3.79	-5.17	-4.60	-3.15	0.29	3.24	6.72	7.72
SACRAMENTO DRAINAGE (SACRAMENTO)														
	RF*	3.73	2.66	2.17	1.54	0.51	0.10	0.01	0.05	0.19	0.99	2.13	3.12	17.22
	ET	0.49	0.71	1.43	2.43	3.78	5.09	6.51	5.65	4.13	2.48	1.12	0.06	34.44
	DIFF	3.24	1.95	0.74	-0.89	-3.27	-4.99	-6.50	-5.60	-3.94	-1.49	1.01	3.06	-17.22
CENTRAL COAST DRAINAGE (SAN FRANCISCO)														
	RF	4.14	3.75	2.84	1.51	0.59	0.13	0.02	0.03	0.18	0.82	1.73	4.02	19.76
	ET	0.83	1.08	1.94	2.75	3.70	4.41	4.99	4.52	3.84	2.76	1.56	1.02	33.40
	DIFF	3.31	2.67	0.90	-1.24	-3.11	-4.28	-4.97	-4.49	-3.66	-1.94	0.17	3.00	-13.64
SOUTH COAST DRAINAGE (LOS ANGELES)														
	RF	3.25	3.55	2.63	1.47	0.32	0.08	0.07	0.15	0.25	0.69	1.27	3.17	16.90
	ET	0.97	1.19	2.12	3.01	4.13	4.94	6.13	5.71	4.56	3.16	1.84	1.20	38.96
	DIFF	2.28	2.36	0.51	-1.54	-3.81	-4.86	-6.06	-5.56	-4.31	-2.47	-0.57	1.97	-22.06

COLORADO		JAN	FEB	MAR	APR	MAY	JUN	JUL	AUG	SEP	OCT	NOV	DEC	TOTAL
PLATTE DRAINAGE BASIN (DENVER)														
	RF	0.50	0.60	1.03	1.90	2.53	1.71	1.77	1.56	1.15	0.95	0.63	0.48	14.81
	ET	0.00	0.00	0.00	1.16	2.25	3.55	4.42	3.89	2.44	1.22	0.00	0.00	18.93
	DIFF	0.50	0.60	1.03	0.74	0.28	-1.84	-2.65	-2.33	-1.29	-0.27	0.63	0.48	-4.12

CONNECTICUT		JAN	FEB	MAR	APR	MAY	JUN	JUL	AUG	SEP	OCT	NOV	DEC	TOTAL
CENTRAL COASTAL (HARTFORD)														
	RF	3.80	3.01	4.27	4.02	3.86	3.62	3.92	4.40	4.14	3.42	4.49	3.88	46.83
	ET	0.00	0.00	0.68	1.80	3.65	5.29	6.37	5.48	3.55	1.96	0.77	0.00	29.55
	DIFF	3.80	3.01	3.59	2.22	0.21	-1.67	-2.45	-1.08	0.59	1.46	3.72	3.88	17.28

DELAWARE		JAN	FEB	MAR	APR	MAY	JUN	JUL	AUG	SEP	OCT	NOV	DEC	TOTAL
NORTHERN (WILMINGTON)														
	RF	3.48	2.85	3.95	3.56	4.00	3.83	4.37	5.24	3.48	3.14	3.48	3.10	44.48
	ET	0.30	0.36	1.04	2.33	4.40	6.18	7.16	6.23	4.28	2.38	1.00	0.39	36.05
	DIFF	3.18	2.49	2.91	1.23	-0.40	-2.35	-2.79	-0.99	-0.80	0.76	2.48	2.71	8.43

*Rainfall data for this location is taken from Kah and Walker (1990).

Selected Evapotranspiration and Rainfall Data (continued)

FLORIDA		JAN	FEB	MAR	APR	MAY	JUN	JUL	AUG	SEP	OCT	NOV	DEC	TOTAL
NORTH (JACKSONVILLE)														
	RF	2.53	3.24	4.02	3.44	3.35	6.13	7.62	6.90	6.55	4.40	1.95	2.52	52.65
	ET	1.34	1.62	2.98	4.49	6.36	7.48	7.87	7.56	6.05	4.27	2.35	1.52	53.89
	DIFF	1.19	1.62	1.04	-1.05	-3.01	-1.35	-0.25	-0.66	0.50	0.13	-0.40	1.00	-1.24
SOUTH CENTRAL (TAMPA)														
	RF	2.09	2.64	3.59	3.38	3.54	7.27	8.14	7.63	7.79	4.14	1.77	1.86	53.84
	ET	1.68	2.04	3.47	4.93	6.66	7.42	7.95	7.56	6.25	4.74	2.86	1.97	57.53
	DIFF	0.41	0.60	0.12	-1.55	-3.12	-0.15	0.19	0.07	1.54	-0.60	-1.09	-0.11	-3.69
LOWER EAST COAST (FT. LAUDERDALE)														
	RF	2.16	2.02	2.82	3.90	5.49	7.49	6.65	6.82	9.47	8.19	2.84	2.15	60.00
	ET	2.03	2.36	3.85	5.22	6.61	7.34	7.86	7.50	6.37	5.08	3.31	2.37	59.90
	DIFF	0.13	-0.34	-1.03	-1.32	-1.12	0.15	-1.21	-0.68	3.10	3.11	-0.47	-0.22	0.10

GEORGIA		JAN	FEB	MAR	APR	MAY	JUN	JUL	AUG	SEP	OCT	NOV	DEC	TOTAL
NORTH CENTRAL (ATLANTA)														
	RF	5.07	5.07	5.57	4.65	3.66	3.75	5.10	4.00	3.23	2.99	3.26	4.96	51.31
	ET	0.67	0.83	1.78	3.27	5.27	6.77	7.33	6.68	4.88	2.96	1.36	0.75	42.55
	DIFF	4.40	4.24	3.79	1.38	-1.61	-3.02	-2.23	-2.68	-1.65	0.03	1.90	4.21	8.76
EAST CENTRAL (AUGUSTA)														
	RF	3.17	3.87	4.34	3.65	3.36	3.76	4.88	4.48	3.80	2.46	2.20	3.46	43.43
	ET	0.95	1.20	2.34	3.93	6.02	7.54	8.10	7.43	5.54	3.48	1.75	1.02	49.30
	DIFF	2.22	2.67	2.00	-0.28	-2.66	-3.78	-3.22	-2.95	-1.74	-1.02	0.45	2.44	-5.87
SOUTHEAST (SAVANNAH)														
	RF	2.50	3.13	3.84	3.19	3.36	5.12	6.83	6.13	6.49	3.25	1.75	2.65	48.24
	ET	1.11	1.39	2.63	4.19	6.20	7.54	7.95	7.46	5.72	3.84	1.99	1.24	51.26
	DIFF	1.39	1.74	1.21	-1.00	-2.84	-2.42	-1.12	-1.33	0.77	-0.59	-0.24	1.41	-3.02

IDAHO		JAN	FEB	MAR	APR	MAY	JUN	JUL	AUG	SEP	OCT	NOV	DEC	TOTAL
SOUTHWESTERN VALLEYS (BOISE)														
	RF	1.54	1.36	1.30	1.04	1.22	0.97	0.21	0.18	0.44	0.91	1.35	1.48	12.00
	ET	0.00	0.34	1.04	2.23	3.83	5.06	6.93	5.89	3.55	1.86	0.61	0.26	31.60
	DIFF	1.54	1.02	0.26	-1.19	-2.61	-4.09	-6.72	-5.71	-3.11	-0.95	0.74	1.22	-19.60

Selected Evapotranspiration and Rainfall Data (continued)

ILLINOIS

		JAN	FEB	MAR	APR	MAY	JUN	JUL	AUG	SEP	OCT	NOV	DEC	TOTAL
NORTHEAST (CHICAGO)														
	RF	1.86	1.58	2.59	3.28	3.75	4.08	3.39	3.38	2.91	2.65	2.09	1.88	33.44
	ET	0.00	0.00	0.68	2.01	3.95	5.89	6.99	6.07	3.87	2.08	0.63	0.00	32.17
	DIFF	1.86	1.58	1.91	1.27	-0.20	-1.81	-3.60	-2.69	-0.96	0.57	1.46	1.88	1.27
WEST SOUTHWEST (SPRINGFIELD)														
	RF	2.01	2.14	2.98	3.68	4.09	4.50	3.26	3.29	2.87	2.95	2.62	1.92	36.31
	ET	0.22	0.35	1.05	2.56	4.60	6.64	7.58	6.62	4.42	2.47	0.83	0.31	37.65
	DIFF	1.79	1.79	1.93	1.12	-0.51	-2.14	-4.32	-3.33	-1.55	0.48	1.79	1.61	-1.34

INDIANA

		JAN	FEB	MAR	APR	MAY	JUN	JUL	AUG	SEP	OCT	NOV	DEC	TOTAL
CENTRAL (INDIANAPOLIS)														
	RF	3.09	2.40	3.42	3.75	4.32	4.39	3.53	3.04	3.33	2.55	3.05	2.52	39.39
	ET	0.00	0.30	0.92	2.32	4.26	6.22	7.12	6.20	4.12	2.27	0.78	0.27	34.78
	DIFF	3.09	2.10	2.50	1.43	0.06	-1.83	-3.59	-3.16	-0.79	0.28	2.27	2.25	4.61

IOWA

		JAN	FEB	MAR	APR	MAY	JUN	JUL	AUG	SEP	OCT	NOV	DEC	TOTAL
CENTRAL (DES MOINES)														
	RF	1.17	1.06	2.08	2.62	4.16	4.94	3.55	3.71	3.18	1.99	1.80	1.10	31.36
	ET	0.00	0.00	0.57	2.01	4.10	5.92	7.03	6.06	3.83	2.05	0.53	0.00	32.10
	DIFF	1.17	1.06	1.51	0.61	0.06	-0.98	-3.48	-2.35	-0.65	-0.06	1.27	1.10	-0.74

KANSAS

		JAN	FEB	MAR	APR	MAY	JUN	JUL	AUG	SEP	OCT	NOV	DEC	TOTAL
NORTHEAST (KANSAS CITY)														
	RF	1.01	1.02	1.99	2.89	4.46	5.40	3.67	4.19	3.66	2.30	1.45	1.18	33.22
	ET	0.00	0.33	0.99	2.63	4.56	6.71	7.77	6.98	4.56	2.58	0.84	0.31	38.26
	DIFF	1.01	0.69	1.00	0.26	-0.10	-1.31	-4.10	-2.79	-0.90	-0.28	0.61	0.87	-5.04
SOUTH CENTRAL (WICHITA)														
	RF	0.74	1.06	1.48	2.35	4.14	3.93	2.96	2.81	2.56	1.95	1.06	0.85	25.89
	ET	0.28	0.47	1.26	2.88	4.87	7.13	8.38	7.58	4.91	2.80	0.96	0.41	41.93
	DIFF	0.46	0.59	0.22	-0.53	-0.73	-3.20	-5.42	-4.77	-2.35	-0.85	0.10	0.44	-16.04

Selected Evapotranspiration and Rainfall Data (continued)

KENTUCKY	JAN	FEB	MAR	APR	MAY	JUN	JUL	AUG	SEP	OCT	NOV	DEC	TOTAL
CENTRAL (LOUISVILLE)													
RF	5.12	3.84	4.99	4.03	3.99	4.32	4.15	3.54	3.03	2.47	3.58	3.85	46.91
ET	0.40	0.55	1.35	2.87	4.80	6.48	7.24	6.52	4.56	2.61	1.03	0.49	38.90
DIFF	4.72	3.29	3.64	1.16	−0.81	−2.16	−3.09	−2.98	−1.53	−0.14	2.55	3.36	8.01

LOUISIANA	JAN	FEB	MAR	APR	MAY	JUN	JUL	AUG	SEP	OCT	NOV	DEC	TOTAL
SOUTHEAST (NEW ORLEANS)													
RF	4.36	4.61	5.20	4.89	4.70	5.52	7.74	6.74	6.64	3.12	3.89	4.99	62.40
ET	1.29	1.60	2.89	4.48	6.37	7.68	8.09	7.65	6.05	4.26	2.25	1.50	54.11
DIFF	3.07	3.01	2.31	0.41	−1.67	−2.16	−0.35	−0.91	0.59	−1.14	1.64	3.49	8.29

MAINE	JAN	FEB	MAR	APR	MAY	JUN	JUL	AUG	SEP	OCT	NOV	DEC	TOTAL
COASTAL (PORTLAND)													
RF	4.22	3.60	3.96	3.65	3.42	3.24	3.07	2.80	3.77	3.68	4.70	3.96	44.07
ET	0.00	0.00	0.46	1.34	2.82	4.17	5.30	4.65	3.03	1.60	0.61	0.00	23.98
DIFF	4.22	3.60	3.50	2.31	0.60	−0.93	−2.23	−1.85	0.74	2.08	4.09	3.96	20.09

MARYLAND	JAN	FEB	MAR	APR	MAY	JUN	JUL	AUG	SEP	OCT	NOV	DEC	TOTAL
NORTHERN CENTRAL (BALTIMORE)													
RF	3.30	2.75	3.86	3.56	4.18	3.87	4.29	4.60	3.71	3.32	3.23	3.19	43.86
ET	0.30	0.39	1.05	2.43	4.41	6.04	6.98	6.19	4.13	2.28	0.94	0.37	35.51
DIFF	3.00	2.36	2.81	1.13	−0.23	−2.17	−2.69	−1.59	−0.42	1.04	2.29	2.82	8.35

MASSACHUSETTS	JAN	FEB	MAR	APR	MAY	JUN	JUL	AUG	SEP	OCT	NOV	DEC	TOTAL
COASTAL (BOSTON)													
RF	4.04	3.37	4.19	3.86	3.23	3.17	2.85	3.85	3.64	3.33	4.11	3.73	43.37
ET	0.00	0.00	0.73	1.71	3.37	4.95	6.18	5.48	3.58	2.05	0.87	0.30	29.22
DIFF	4.04	3.37	3.46	2.15	−0.14	−1.78	−3.33	−1.63	0.06	1.28	3.24	3.43	14.15

Selected Evapotranspiration and Rainfall Data (continued)

MICHIGAN

	JAN	FEB	MAR	APR	MAY	JUN	JUL	AUG	SEP	OCT	NOV	DEC	TOTAL
SOUTHEAST LOWER (DETROIT)													
RF	1.89	1.90	2.29	2.94	3.38	3.29	2.75	2.91	2.58	2.61	2.21	1.93	30.68
ET	0.00	0.00	0.57	1.78	3.63	5.52	6.56	5.68	3.68	1.96	0.63	0.00	30.01
DIFF	1.89	1.90	1.72	1.16	-0.25	-2.23	-3.81	-2.77	-1.10	0.65	1.58	1.93	0.67

MINNESOTA

	JAN	FEB	MAR	APR	MAY	JUN	JUL	AUG	SEP	OCT	NOV	DEC	TOTAL
EAST CENTRAL (MINNEAPOLIS)													
RF	0.92	0.88	1.78	2.28	3.82	4.73	3.48	3.98	3.15	1.77	1.68	0.93	29.40
ET	0.00	0.00	0.00	1.70	3.71	5.58	6.66	5.71	3.42	1.70	0.37	0.00	28.85
DIFF	0.92	0.88	1.78	0.58	0.11	-0.85	-3.18	-1.73	-0.27	0.07	1.31	0.93	0.55

MISSISSIPPI

	JAN	FEB	MAR	APR	MAY	JUN	JUL	AUG	SEP	OCT	NOV	DEC	TOTAL
COASTAL (BILOXI)													
RF	4.31	4.59	6.23	5.38	4.82	5.17	7.81	5.94	5.90	2.67	3.71	5.21	61.74
ET	1.12	1.45	2.67	4.21	6.16	7.49	8.02	7.56	5.84	3.91	2.01	1.32	51.76
DIFF	3.19	3.14	3.56	1.17	-1.34	-2.32	-0.21	-1.62	0.06	-1.24	1.70	3.89	9.98
SOUTHWEST (VICKSBURG)													
RF	5.44	5.21	5.60	5.15	4.77	4.17	5.14	3.70	2.99	2.32	4.21	5.59	54.29
ET	0.95	1.26	2.44	4.03	6.00	7.52	8.11	7.46	5.66	3.60	1.76	1.12	49.91
DIFF	4.49	3.95	3.16	1.12	-1.23	-3.35	-2.97	-3.76	-2.67	-1.28	2.45	4.47	4.38

MISSOURI

	JAN	FEB	MAR	APR	MAY	JUN	JUL	AUG	SEP	OCT	NOV	DEC	TOTAL
NORTHEAST PRAIRIE (ST. LOUIS)													
RF	1.40	1.37	2.42	3.21	4.50	5.58	3.49	4.13	3.43	2.67	1.80	1.50	35.50
ET	0.00	0.29	0.97	2.54	4.57	6.59	7.77	6.79	4.40	2.47	0.78	0.27	37.44
DIFF	1.40	1.08	1.45	0.67	-0.07	-1.01	-4.28	-2.66	-0.97	0.20	1.02	1.23	-1.94
NORTHWEST PRAIRIE (KANSAS CITY)													
RF	1.83	1.85	2.88	3.58	4.10	4.72	3.56	3.51	3.17	2.96	2.35	1.85	36.36
ET	0.00	0.33	1.04	2.56	4.60	6.62	7.60	6.78	4.42	2.49	0.83	0.30	37.57
DIFF	1.83	1.52	1.84	1.02	-0.50	-1.90	-4.04	-3.27	-1.25	0.47	1.52	1.55	-1.21

Selected Evapotranspiration and Rainfall Data (continued)

MONTANA

WESTERN (MISSOULA)

	JAN	FEB	MAR	APR	MAY	JUN	JUL	AUG	SEP	OCT	NOV	DEC	TOTAL
RF	1.81	1.47	1.35	1.25	1.82	2.26	0.99	0.96	1.29	1.63	1.87	1.96	18.66
ET	0.00	0.00	0.51	1.45	2.86	3.91	5.22	4.38	2.56	1.25	0.32	0.00	22.46
DIFF	1.81	1.47	0.84	-0.20	-1.04	-1.65	-4.23	-3.42	-1.27	0.38	1.55	1.96	-3.80

NEBRASKA

EAST CENTRAL (OMAHA)

	JAN	FEB	MAR	APR	MAY	JUN	JUL	AUG	SEP	OCT	NOV	DEC	TOTAL
RF	0.83	0.96	1.51	2.43	3.70	4.55	3.17	3.49	2.62	1.40	1.09	0.77	26.52
ET	0.00	0.00	0.73	2.22	4.25	6.25	7.60	6.62	4.13	2.25	0.63	0.00	34.68
DIFF	0.83	0.96	0.78	0.21	-0.55	-1.70	-4.43	-3.13	-1.51	-0.85	0.46	0.77	-8.16

NEVADA

NORTHWESTERN (RENO)

	JAN	FEB	MAR	APR	MAY	JUN	JUL	AUG	SEP	OCT	NOV	DEC	TOTAL
RF	1.20	1.12	0.81	0.65	0.74	0.58	0.25	0.15	0.28	0.65	0.74	1.13	8.30
ET	0.00	0.37	0.93	1.88	3.25	4.48	6.21	5.37	3.42	1.81	0.64	0.31	28.67
DIFF	1.20	0.75	-0.12	-1.23	-2.51	-3.90	-5.96	-5.22	-3.14	-1.16	0.10	0.82	-20.37

EXTREME SOUTHERN (LAS VEGAS)

	JAN	FEB	MAR	APR	MAY	JUN	JUL	AUG	SEP	OCT	NOV	DEC	TOTAL
RF	0.61	0.63	0.49	0.35	0.18	0.10	0.49	0.52	0.38	0.37	0.41	0.57	5.10
ET	0.62	0.91	2.03	3.55	5.57	7.50	9.24	8.15	5.71	3.30	1.41	0.78	48.77
DIFF	-0.01	-0.28	-1.54	-3.20	-5.39	-7.40	-8.75	-7.63	-5.33	-2.93	-1.00	-0.21	-43.67

NEW HAMPSHIRE

SOUTHERN (CONCORD)

	JAN	FEB	MAR	APR	MAY	JUN	JUL	AUG	SEP	OCT	NOV	DEC	TOTAL
RF	3.44	2.69	3.48	3.52	3.53	3.60	3.71	3.18	3.81	3.01	3.90	3.35	41.22
ET	0.00	0.00	0.46	1.50	3.26	4.88	5.89	5.01	3.15	1.60	0.56	0.00	26.31
DIFF	3.44	2.69	3.02	2.02	0.27	-1.28	-2.18	-1.83	0.66	1.41	3.34	3.35	14.91

NEW JERSEY

NORTHERN (NEWARK)

	JAN	FEB	MAR	APR	MAY	JUN	JUL	AUG	SEP	OCT	NOV	DEC	TOTAL
RF	3.47	2.92	4.10	3.91	4.07	4.01	4.65	4.91	4.06	3.42	3.88	3.56	46.96
ET	0.00	0.26	0.85	2.10	4.04	5.65	6.72	5.83	3.84	2.15	0.87	0.28	32.59
DIFF	3.47	2.66	3.25	1.81	0.03	-1.64	-2.07	-0.92	0.22	1.27	3.01	3.28	14.37

Selected Evapotranspiration and Rainfall Data (continued)

NEW MEXICO	JAN	FEB	MAR	APR	MAY	JUN	JUL	AUG	SEP	OCT	NOV	DEC	TOTAL
NORTHERN MOUNTAINS (SANTA FE)													
RF	0.81	0.80	0.95	1.12	1.65	1.24	2.39	2.76	1.64	1.28	0.70	0.73	16.07
ET	0.00	0.30	0.68	1.56	2.82	4.26	5.05	4.51	3.02	1.63	0.52	0.00	24.35
DIFF	0.81	0.50	0.27	-0.44	-1.17	-3.02	-2.66	-1.75	-1.38	-0.35	0.18	0.73	-8.28
CENTRAL VALLEY (ALBUQUERQUE)													
RF	0.41	0.44	0.41	0.45	0.61	0.72	1.46	1.61	1.20	0.81	0.32	0.47	8.91
ET	0.38	0.64	1.44	2.76	4.58	6.37	7.17	6.43	4.42	2.52	0.93	0.46	38.10
DIFF	0.03	-0.20	-1.03	-2.31	-3.97	-5.65	-5.71	-4.82	-3.22	-1.71	-0.61	0.01	-29.19

NEW YORK	JAN	FEB	MAR	APR	MAY	JUN	JUL	AUG	SEP	OCT	NOV	DEC	TOTAL
COASTAL (NEW YORK)													
RF	3.64	3.14	4.37	3.75	3.71	3.39	3.78	4.68	3.91	3.46	3.97	3.67	45.47
ET	0.25	0.30	0.86	2.01	3.91	5.65	6.89	6.02	4.09	2.36	0.99	0.37	33.70
DIFF	3.39	2.84	3.51	1.74	-0.20	-2.26	-3.11	-1.34	-0.18	1.10	2.98	3.30	11.77
GREAT LAKES (BUFFALO)													
RF	2.67	2.49	2.93	2.97	3.16	2.76	2.94	2.88	3.15	2.99	3.06	2.83	34.83
ET	0.00	0.00	0.51	1.60	3.37	5.17	6.19	5.47	3.55	1.88	0.68	0.00	28.42
DIFF	2.67	2.49	2.42	1.37	-0.21	-2.41	-3.25	-2.59	-0.40	1.11	2.38	2.83	6.41

NORTH CAROLINA	JAN	FEB	MAR	APR	MAY	JUN	JUL	AUG	SEP	OCT	NOV	DEC	TOTAL
SOUTHERN MOUNTAINS (ASHEVILLE)													
RF	8.83	4.71	5.49	4.45	4.00	4.47	5.94	5.45	3.84	3.60	3.54	4.71	59.03
ET	0.50	0.63	1.35	2.65	4.33	5.83	6.36	5.76	4.11	2.40	1.03	0.56	35.51
DIFF	8.33	4.08	4.14	1.80	-0.33	-1.36	-0.42	-0.31	-0.27	1.20	2.51	4.15	23.52
CENTRAL PIEDMONT (RALEIGH)													
RF	3.73	3.66	4.17	3.71	3.71	3.79	5.60	5.07	4.00	3.01	2.99	3.52	46.96
ET	0.59	0.77	1.68	3.27	5.20	6.88	7.60	6.73	4.86	2.85	1.29	0.65	42.37
DIFF	3.14	2.89	2.49	0.44	-1.49	-3.09	-2.00	-1.66	-0.86	0.16	1.70	2.87	4.59

Selected Evapotranspiration and Rainfall Data (continued)

NORTH DAKOTA

SOUTH CENTRAL (BISMARK)

	JAN	FEB	MAR	APR	MAY	JUN	JUL	AUG	SEP	OCT	NOV	DEC	TOTAL
RF	0.43	0.45	0.76	1.24	2.19	3.80	2.19	1.88	1.28	0.96	0.57	0.32	16.07
ET	0.00	0.00	0.00	1.45	3.40	4.93	6.55	5.57	3.06	1.42	0.00	0.00	26.38
DIFF	0.43	0.45	0.76	-0.21	-1.21	-1.13	-4.36	-3.69	-1.78	-0.46	0.57	0.32	-10.31

OHIO

NORTHEAST (CLEVELAND)

	JAN	FEB	MAR	APR	MAY	JUN	JUL	AUG	SEP	OCT	NOV	DEC	TOTAL
RF	2.77	2.31	3.12	3.47	3.68	3.74	3.60	2.32	2.96	2.74	2.70	2.40	36.81
ET	0.00	0.00	0.68	1.88	3.74	5.59	6.34	5.63	3.81	2.07	0.74	0.25	30.73
DIFF	2.77	2.31	2.44	1.59	-0.06	-1.85	-2.74	-3.31	-0.85	0.67	1.96	2.15	6.08

SOUTHWEST (DAYTON)

	JAN	FEB	MAR	APR	MAY	JUN	JUL	AUG	SEP	OCT	NOV	DEC	TOTAL
RF	3.59	2.80	3.79	3.70	3.87	4.23	3.84	3.20	2.95	2.32	2.95	2.74	39.98
ET	0.27	0.36	1.04	2.43	4.40	6.25	7.14	6.22	4.25	2.36	0.87	0.34	35.93
DIFF	3.32	2.44	2.75	1.27	-0.53	-2.02	-3.30	-3.02	-1.30	-0.04	2.08	2.40	4.05

OKLAHOMA

CENTRAL (OKLAHOMA CITY)

	JAN	FEB	MAR	APR	MAY	JUN	JUL	AUG	SEP	OCT	NOV	DEC	TOTAL
RF	1.43	1.58	2.08	3.44	5.44	4.46	3.07	2.69	3.35	2.93	1.81	1.53	33.81
ET	0.47	0.72	1.69	3.41	5.36	7.32	8.51	7.92	5.34	3.18	1.27	0.64	45.83
DIFF	0.96	0.86	0.39	0.03	0.08	-2.86	-5.44	-5.23	-1.99	-0.25	0.54	0.89	-12.02

OREGON

WILLAMETTE VALLEY (PORTLAND)

	JAN	FEB	MAR	APR	MAY	JUN	JUL	AUG	SEP	OCT	NOV	DEC	TOTAL
RF	7.71	6.17	5.88	3.29	2.78	2.15	0.50	0.72	1.87	4.85	7.25	8.62	51.79
ET	0.41	0.64	1.27	2.21	3.39	4.17	5.23	4.65	3.33	1.96	0.90	0.55	28.71
DIFF	7.30	5.53	4.61	1.08	-0.61	-2.02	-4.73	-3.93	-1.46	2.89	6.35	8.07	23.08

PENNSYLVANIA

SOUTHEASTERN PIEDMONT (PHILADELPHIA)

	JAN	FEB	MAR	APR	MAY	JUN	JUL	AUG	SEP	OCT	NOV	DEC	TOTAL
RF	3.21	2.67	3.88	3.51	3.99	3.90	4.40	4.66	3.52	3.16	3.48	3.15	43.53
ET	0.25	0.32	0.97	2.31	4.38	6.03	7.13	6.18	4.12	2.28	0.93	0.34	35.24
DIFF	2.96	2.35	2.91	1.20	-0.39	-2.13	-2.73	-1.52	-0.60	0.88	2.55	2.81	8.29

Selected Evapotranspiration and Rainfall Data (continued)

RHODE ISLAND	JAN	FEB	MAR	APR	MAY	JUN	JUL	AUG	SEP	OCT	NOV	DEC	TOTAL
RHODE ISLAND (COMPLETE STATE)													
RF	3.96	3.26	4.24	3.86	3.25	2.95	2.85	4.14	3.54	3.19	4.32	3.72	43.28
ET	0.00	0.26	0.73	1.78	3.47	4.97	6.20	5.51	3.68	2.07	0.87	0.30	29.84
DIFF	3.96	3.00	3.51	2.08	-0.22	-2.02	-3.35	-1.37	-0.14	1.12	3.45	3.42	13.44

SOUTH CAROLINA	JAN	FEB	MAR	APR	MAY	JUN	JUL	AUG	SEP	OCT	NOV	DEC	TOTAL
SOUTHERN (CHARLESTON)													
RF	2.57	3.35	3.89	3.00	3.56	4.67	6.54	6.05	5.23	2.71	2.13	2.81	46.51
ET	0.96	1.20	2.35	3.93	6.00	7.36	7.91	7.27	5.54	3.58	1.82	1.07	48.99
DIFF	1.61	2.15	1.54	-0.93	-2.44	-2.69	-1.37	-1.22	-0.31	-0.87	0.31	1.74	-2.48

SOUTH DAKOTA	JAN	FEB	MAR	APR	MAY	JUN	JUL	AUG	SEP	OCT	NOV	DEC	TOTAL
SOUTHEAST (SIOUX FALLS)													
RF	0.55	0.61	0.98	2.10	2.57	4.02	2.48	2.50	1.57	1.20	0.75	0.52	19.85
ET	0.00	0.00	0.00	1.61	3.61	5.45	6.81	5.94	3.32	1.59	0.29	0.00	28.62
DIFF	0.55	0.61	0.98	0.49	-1.04	-1.43	-4.33	-3.44	-1.75	-0.39	0.46	0.52	-8.77

TENNESSEE	JAN	FEB	MAR	APR	MAY	JUN	JUL	AUG	SEP	OCT	NOV	DEC	TOTAL
WESTERN (MEMPHIS)													
RF	5.96	4.65	5.28	4.42	4.16	3.96	4.04	3.05	3.30	2.85	4.27	4.47	50.41
ET	0.51	0.72	1.68	3.37	5.36	7.13	7.88	7.28	5.02	2.96	1.22	0.64	43.77
DIFF	5.45	3.93	3.60	1.05	-1.20	-3.17	-3.84	-4.23	-1.72	-0.11	3.05	3.83	6.64

Selected Evapotranspiration and Rainfall Data (continued)

TEXAS		JAN	FEB	MAR	APR	MAY	JUN	JUL	AUG	SEP	OCT	NOV	DEC	TOTAL
NORTH CENTRAL (DALLAS)														
	RF	2.13	2.35	2.14	3.70	4.78	3.18	2.18	1.93	2.72	2.87	2.21	2.31	32.50
	ET	0.73	1.04	2.23	3.91	5.96	7.77	8.79	8.29	5.84	3.71	1.66	0.95	50.88
	DIFF	1.40	1.31	-0.09	-0.21	-1.18	-4.59	-6.61	-6.36	-3.12	-0.84	0.55	1.36	-18.38
SOUTH CENTRAL (SAN ANTONIO)														
	RF	2.18	2.36	1.98	3.05	3.69	3.09	2.49	2.57	3.63	2.86	2.15	2.42	32.47
	ET	1.17	1.53	2.90	4.63	6.56	7.94	8.70	8.23	6.21	4.30	2.18	1.39	55.74
	DIFF	1.01	0.83	-0.92	-1.58	-2.87	-4.85	-6.21	-5.66	-2.58	-1.44	-0.03	1.03	-23.27
UPPER COAST (HOUSTON)														
	RF	3.41	3.60	2.67	3.39	4.16	3.55	4.29	4.51	4.40	3.64	3.43	3.89	44.94
	ET	1.19	1.54	2.88	4.50	6.51	7.87	8.32	7.85	6.06	4.28	2.19	1.44	54.63
	DIFF	2.22	2.06	-0.21	-1.11	-2.35	-4.32	-4.03	-3.34	-1.66	-0.64	1.24	2.45	-9.69

UTAH		JAN	FEB	MAR	APR	MAY	JUN	JUL	AUG	SEP	OCT	NOV	DEC	TOTAL
NORTH CENTRAL (SALT LAKE CITY)														
	RF	1.57	1.44	1.66	1.81	1.60	1.09	0.61	0.84	0.73	1.37	1.43	1.52	15.67
	ET	0.00	0.29	0.86	2.01	3.59	5.07	6.92	6.01	3.70	1.90	0.59	0.24	31.18
	DIFF	1.57	1.15	0.80	-0.20	-1.99	-3.98	-6.31	-5.17	-2.97	-0.53	0.84	1.28	-15.51

VERMONT		JAN	FEB	MAR	APR	MAY	JUN	JUL	AUG	SEP	OCT	NOV	DEC	TOTAL
NORTHEASTERN (MONTPELIER)														
	RF	2.49	2.19	2.49	2.98	3.37	3.83	3.95	3.37	3.69	3.25	3.25	2.58	37.44
	ET	0.00	0.00	0.00	1.25	3.06	4.65	5.48	4.65	2.83	1.42	0.44	0.00	23.78
	DIFF	2.49	2.19	2.49	1.73	0.31	-0.82	-1.53	-1.28	0.86	1.83	2.81	2.58	13.66

VIRGINIA		JAN	FEB	MAR	APR	MAY	JUN	JUL	AUG	SEP	OCT	NOV	DEC	TOTAL
EASTERN PIEDMONT (RICHMOND)														
	RF	3.51	2.97	3.71	3.53	3.69	3.67	5.09	4.78	3.64	2.92	3.00	3.09	43.60
	ET	0.47	0.59	1.42	2.88	4.83	6.45	7.24	6.50	4.55	2.60	1.15	0.53	39.21
	DIFF	3.04	2.38	2.29	0.65	-1.14	-2.78	-2.15	-1.72	-0.91	0.32	1.85	2.56	4.39

Selected Evapotranspiration and Rainfall Data (continued)

WASHINGTON

PUGET SOUND LOWLAND (SEATTLE)

	JAN	FEB	MAR	APR	MAY	JUN	JUL	AUG	SEP	OCT	NOV	DEC	TOTAL
RF	5.59	4.44	3.96	2.54	2.00	1.89	0.89	1.01	2.06	4.11	5.58	6.32	40.39
ET	0.39	0.59	1.20	2.14	3.35	4.26	5.02	4.40	3.13	1.83	0.82	0.52	27.65
DIFF	5.20	3.85	2.76	0.40	-1.35	-2.37	-4.13	-3.39	-1.07	2.28	4.76	5.80	12.74

WEST VIRGINIA

SOUTHWESTERN (CHARLESTON)

	JAN	FEB	MAR	APR	MAY	JUN	JUL	AUG	SEP	OCT	NOV	DEC	TOTAL
RF	3.78	3.26	4.19	3.47	3.85	4.14	4.93	3.80	2.90	2.29	2.74	3.02	42.37
ET	0.37	0.47	1.19	2.66	4.67	6.31	7.07	6.34	4.41	2.47	0.96	0.44	37.36
DIFF	3.41	2.79	3.00	0.81	-0.82	-2.17	-2.14	-2.54	-1.51	-0.18	1.78	2.58	5.01

WISCONSIN

SOUTHEAST (MILWAUKEE)

	JAN	FEB	MAR	APR	MAY	JUN	JUL	AUG	SEP	OCT	NOV	DEC	TOTAL
RF	1.75	1.29	2.49	2.88	3.39	3.99	3.66	3.25	2.85	2.11	2.22	1.68	31.56
ET	0.00	0.00	0.47	1.71	3.52	5.39	6.49	5.74	3.58	1.88	0.52	0.00	29.30
DIFF	1.75	1.29	2.02	1.17	-0.13	-1.40	-2.83	-2.49	-0.73	0.23	1.70	1.68	2.26

WYOMING

PLATTE DRAINAGE (CHEYENNE)

	JAN	FEB	MAR	APR	MAY	JUN	JUL	AUG	SEP	OCT	NOV	DEC	TOTAL
RF	0.54	0.58	0.95	1.58	2.04	1.79	1.40	1.10	1.00	0.86	0.62	0.54	13.00
ET	0.00	0.00	0.42	1.31	2.64	4.21	5.59	4.80	2.81	1.43	0.37	0.00	23.58
DIFF	0.54	0.58	0.53	0.27	-0.60	-2.42	-4.19	-3.70	-1.81	-0.57	0.25	0.54	-10.58

Appendix B

Water Measurements, Factors, and Formulas for Calculating Water Requirements, Water Budgets, and Irrigation Schedules

Water Measurements:

1 gallon	=	8.34 pounds		
1 cubic foot	=	7.48 gallons		
1 unit (Ccf)	=	748 gallons	=	100 cubic feet
1 acre foot	=	325,580 gallons	=	43,560 cubic feet
1 second foot	=	7.48 gallons/second	=	1 cubic foot/second

Meter Capacity:

⅝-in. disc	=	20 gallons per minute
¾-in. disc	=	30 gallons per minute
1-in. disc	=	50 gallons per minute
1 ½-in. disc or turbine	=	100 gallons per minute
2-in. disc, turbine or compound	=	160 gallons per minute
3-in. disc	=	300 gallons per minute
3-in. turbine	=	320 gallons per minute
3-in. compound	=	350 gallons per minute
4-in. disc or compound	=	500 gallons per minute
4-in. turbine	=	600 gallons per minute
6-in. disc or compound	=	1000 gallons per minute
6-in. turbine	=	1.250 gallons per minute

Conversion Factors:

Acres to Square Feet
   (43,560 square feet/acre) (acres) = square feet

Square Feet to Acres
   (square feet)/(43,560 square foot/acre) = acres

Inches of Water to Gallons
   (0.623 gallons/square foot inch) (inches of water) = gallons/square feet

Inches of Water to 100 Cubic Feet of Water
(0.00083 Ccf of water/square foot inch) (inches of water)
= Ccf of water/square feet

0.00083 is the decimal fraction of 100 cubic feet (Ccf) that will cover one square foot with water one inch deep.

Irrigation Efficiency:

As a percentage of total water use:

IE = [(amount of water beneficially applied)/(total amount of water applied)] × 100

As a decimal fraction of hardware and management efficiency:

IE = (hardware efficiency) (management efficiency)

ET of a Specific Crop:

$$ET_c = ET_O \times K_c$$

Where:

ET_c = ET rate of a specific crop
ET_O = reference evapotranspiration
K_c = crop coefficient

Landscape Coefficient:

$$k_l = (k_s)(k_d)(k_{mc})$$

Where:

k_l = landscape coefficient
k_s = species factor
k_d = density factor
k_{mc} = microclimate factor

See "Estimating Water Requirements of Landscape Plantings," Cooperative Extension, University of California Division of Agriculture and Natural Resources Leaflet 21493.

Plant Factor:

Decimal Fraction of ET_O:

High water use	=	0.5 to 0.8
Moderate water use	=	0.3 to 0.5
Low water use	=	0.0 to 0.3

Weighted Plant Factor:
(decimal fraction of ET_O) (decimal fraction of landscape area)
= weighted factor

Average Plant Factor:
average plant factor = sum of weighted factors

ET Adjustment Factor:

ET_O adjustment factor = average plant factor/irrigation efficiency

Water Budget Formulas:

Simple Water Budget Formula:

(ET_O) (area) (conversion factor) = water budget

Where:

ET_O = local reference evapotranspiration
area = square feet of landscape or irrigated area
conversion factor = constant used to obtain desired units of measure

Water Budget in Gallons:

(inches of ET) (square feet of area) (0.623 gallons/square feet inches) = gallons

Water Budget in Ccf:

(inches of ET) (square feet of area) (0.00083 Ccf/square feet inches) = Ccf

Adjusted Water Budget:

(ET_O) (adjustment factor) (area) (conversion factor) = adjusted water budget

Where:

ET_O = local reference evapotranspiration

To calculate an annual budget, use annual ET_O; for monthly budgets, use monthly ET_O.

Effective Precipitation (Beneficial Rainfall):

Effective precipitation ≤ precipitation

and

Effective precipitation ≤ accumulated daily depletion due to ET losses

Cumulative depletion = accumulated daily depletion − effective precipitation

Examples:

Given weekly rainfall (precipitation) of 1 in. and weekly ET (accumulated daily depletion) of 2 in. for the same period: precipitation is less than ET (1 in. ≤ 2 in.). Therefore, effective precipitation is 1 in. Cumulative depletion, the amount of water needed from supplemental irrigation, is the accumulated daily depletion minus effective precipitation or 2 in. − 1 in. = 1 in.

Given weekly rainfall (precipitation) of 3 in. and weekly ET (accumulated daily depletion) of 1 in. for the same period: precipitation is greater than ET (3 in. ≥ 1 in.). Since effective precipitation must be less than or equal to accumulated daily depletion, effective precipitation is 1 in. In this case, no supplemental irrigation is required because the cumulative depletion is less than the effective precipitation, or 1 in. − 3 in. = (−2 in.).

Pooling:

Looped Pooling:

Water allocations for pooled accounts are calculated by adding the assigned allocation for all accounts to yield a pooled allocation. Each billing period, the consumption for all the accounts is also totaled to yield pooled consumption and then compared to the pooled allotment to determine if excess use has occurred:

pooled allocation = sum of assigned allocations
pooled consumption = sum of consumption
excess use = pooled allocation − pooled consumption

Pooled allocations and consumption can be calculated in either gallons per day or billing units. A rate structure or water service contingencies serve as enforcement mechanisms.

Factored Pooling:

To calculate allocations by factored pooling, the assigned allocations for all the accounts to be pooled are totaled. The landscape manager must monitor current consumption, then project the rate of use for each account. The projected rates of use for all accounts are totaled, then divided by the total assigned allocations:

total assigned allocations = sum of gallons per day allocations for pooled accounts
total projected rate of use = sum of gallons per day projected rates of use for pooled accounts
(total projected rate of use)/(total assigned allocations) = pooling factor

This calculation yields a factor that is used to adjust the assigned allocations for the individual accounts. The projected rate of use for each account is divided by the pooling factor to calculate the pooled allocation for each account:

(projected rate of use)/(pooling factor) = pooled allocation

This method is more complex than looped pooling but requires the customer to establish consumption goals. The calculations should be performed by the customer and submitted for review before the end of the billing period to which the calculations apply. Accounts that are pooled together should be in the same customer name or under the same management and should have comparable billing periods. Incentive to pool in this manner depends on a rate structure that includes excess use charges.

Precipitation Rate:

General Formula:

$$PR = (96.3)(gpm)/(area)$$

Where:

PR = the average precipitation rate in inches per hour
96.3 = a constant, which incorporates inches per square foot per hour
gpm = the total gallons per minute applied to the area by sprinklers
area = area irrigated by a circuit in square feet

Square Spacing Sprinkler Pattern:

$$PR = (96.3)(gpm)/(spacing)^2$$

Where:

gpm = the flow rate of a full circle sprinkler
spacing = the spacing between sprinklers

Triangular Spacing Sprinkler Pattern:

$$PR = (96.3)(gpm)/(spacing)^2 (0.866)$$

Where:

gpm = the flow rate of a full-circle sprinkler
0.866 = a constant used to calculate the area of coverage

Estimated Water-Holding Capacity:

Soil Texture	Total Water-Holding Capacity Per Foot of Soil Depth*	Average Allowable Depletion Per Foot of Soil Depth*
Sand	0.6–1.8	0.75
Sandy loam	1.8–2.7	1.1
Loam	2.7–4.0	1.7
Silt loam	4.0–4.7	2.1
Clay loam	4.2–4.7	2.0
Clay	4.5–4.9	2.0

Source: Irrigation Management Group (1992)
*Inches of water; an inch of water covers the surface 1 in. deep.

Allowable Depletion (AD):

allowable depletion (AD) = (average allowable depletion) (root depth)

Where:

average allowable depletion = allowable depletion per foot of soil depth (See estimate from chart, p. 83.)

root depth = approximate rooting depth of plants under irrigation in feet

Estimated Daily ET Rate:

estimated ET = (seasonal, monthly, or weekly ET_O) (plant factor)/(days in season)

Where:

estimated ET = daily ET rate based on historic averages

ET_O = historic average reference ET for a season, month, or week

plant factor = decimal fraction of reference ET for plants served by an irrigation circuit for which schedule is being calculated. The highest plant factor should be used when plants of mixed requirements are irrigated together. A "landscape coefficient" will provide a more accurate measure of plant water requirements and can be used where the factors that comprise the landscape coefficient are known. (See Plant Factor and Landscape Coefficient above.)

days in season = corresponding number of days in the period represented by ET_O

Irrigation Frequency:

irrigation frequency = AD/estimated ET

Where:

irrigation frequency = the interval in days between water applications. An application may need to be divided into more than one cycle to prevent runoff

AD = allowable depletion

estimated ET = estimated daily ET rate

Irrigation frequency should be rounded to the nearest even number.

Water Requirement:

$$\text{water requirement} = \text{AD/hardware efficiency}$$

Where:

water requirement	=	volume of a water application in inches
AD	=	allowable depletion
hardware efficiency	=	a percent or decimal fraction of water beneficially applied (See chapter 4.)

Duration:

$$\text{duration} = \text{IR/PR}$$

Where:

duration	=	the length of an irrigation cycle in hours
IR	=	infiltration rate in inches per hour (See Table 5-1.)
PR	=	precipitation rate in inches per hour

Number of Cycles:

$$\text{number of cycles} = \text{water requirement/IR}$$

Where:

number of cycles	=	the number of times a circuit must operate per day to satisfy the water requirement
water requirement	=	See above.
IR	=	infiltration rate

Metric Conversions:

surface area
- acres × 0.4047 = hectares
- acres × 4047 = square metres
- square feet × 0.0929 = square metres

evapotranspiration, measure of water required by plants
- inches × 2.540 = centimetres
- inches × 25.40 = millimetres
- gallons × 3.785 = litres

application rate/infiltration rate
- inches per hour × 2.540 = centimetres per hour

lawn area (linear distance)
- feet × 0.305 = metres

flow rate from irrigation unit
    gallons per minute × 0.2271 = cubic metres per hour

delivery of water
    gallons per hour × 3.785 = litres per hour

cutting height for grasses/depth of mulch
    inches × 2.540 = centimetres

others
    cubic feet × 0.02832 = cubic metres
    acre feet × (1.233 × 10^6) = litres
    acre feet × 1233 = cubic metres

metric water budget formula

$$\frac{\text{(millimetres of ET)(square metres of area)}}{\text{(1 litre/square metre millimetre)}} = \text{litres}$$

NOTE:

It takes 1 litre of water to cover one square metre with one millimetre of water. For an adjusted metric water budget formula, include an adjustment factor that is a decimal fraction of ET.

Appendix C

Examples of Landscape Standards

Sample Checklist Approach:
Marin Municipal Water District
Ordinance No. 326
Corte Madera, Calif.

Sample Combined Checklist and Water Budget Approach:
Landscape Water Conservation Ordinance
Modeled After California State Ordinance AB325
Hayward, Calif.

Sample Water Budget Approach:
East Bay Municipal Utility District
Irrigation Incentive Program
Oakland, Calif.

Sample Checklist Approach:

Marin Municipal Water District
Ordinance No. 326

Features:

- public water utility as the implementing agency
- specific criteria for landscape design and installation including percent limitations on turfgrass and landscape elements that use a lot of water
- plan review, certification, and site inspection process that involves landscape professionals as signatory
- checklist, submittal, and certification forms

Implementing Agency:
Marin Municipal Utility District
220 Nellen Avenue
Corte Madera, CA 94925-1169
(415) 924-4600

APPENDIX C 105

MARIN MUNICIPAL WATER DISTRICT

ORDINANCE NO. 326

AN ORDINANCE REVISING WATER CONSERVATION REQUIREMENTS

BE IT ORDAINED BY THE BOARD OF DIRECTORS OF THE MARIN MUNICIPAL WATER DISTRICT AS FOLLOWS:

Section 1. Section 11.04.080 of the Marin Municipal Water District (MMWD) Code is repealed and Chapter 11.60 is added to Title 11 of the District's code to read as follows:

11.60.010 **Purpose.** All applicants for new, increased, or modified service shall comply with the requirements set forth in this chapter in addition to those set forth in Chapter 11.04 of this Code as a condition of receiving service.

11.60.020 **Definitions.** Definitions used in this chapter are as follows:

(1) Developed landscape area: All outdoor areas under irrigation, surrounding hardscape areas, swimming pools, and decorative pools and fountains.

(2) Hardscape: Patios, decks and paths. Does not include driveways and sidewalks.

(3) High-water-use plants: Annuals, plants in containers, and plants not on MMWD's list of low-water-use plants.

(4) Landscape Plans: This includes a planting plan, an irrigation plan, and a grading plan. All plans must be drawn at a scale that clearly and accurately identifies plants, irrigation layout, equipment, and finish grades. Landscape plans shall include the following:

(A) Planting Plan: Planting plans must accurately identify and locate, but are not limited to the following:

(i) New and existing trees, shrubs, groundcovers and turf areas within the developed landscape area;

(ii) Plants by botanical name, common name, container size, spacing and quantities;

(iii) Property lines, streets and street names;

(iv) Driveway(s), sidewalk(s) and other hardscape features as necessary;

(v) Pool(s), fountain(s), fence(s) and retaining wall(s);

(vi) Existing and proposed buildings;

(vii) The square footage(s) of the various landscape hydrozones on the plan. Hydrozones are separate portions of the landscape area having plants with similar water needs that are served by a valve or set of valves with the same setting. These hydrozones include, but are not limited to turf, high-water-use plants with overhead irrigation, low-water-use plants with overhead irrigation, drip irrigation and fountain and pool areas. If more than one water meter serves the site, the individual hydrozones must be identified with the meter providing water service.

(B) Irrigation Plan: The irrigation plan shall be drawn at the same scale as the planting plan. The irrigation plan will be separate from but in the same format as the planting plan. The irrigation plan shall show but not be limited to the following:

(i) Irrigation system point of connection;

(ii) Water service pressure at point of irrigation system connection;

(iii) Water meter size;

(iv) Backflow prevention devices;

(v) Major components of the irrigation system;

(vi) Total precipitation rate in inches per hour for each overhead irrigation circuit;

(vii) Total flow rate (gph) and operating pressure (psi) for each irrigation circuit;

(viii) Irrigation legend will have the following elements: Symbols for various irrigation equipment, general description of equipment, manufacture name and model number, operating pressure, manufacturer's irrigation nozzle rating in gallons per minute (gpm) or gallons per hour (gph) as necessary, minimum and maximum spray radius, manufacturer's rated precipitation rate per nozzle;

(ix) Reclaimed water piping and guidelines as required.

(C) Grading Plan: The grading plan shall be drawn at the same scale as the planting and irrigation plan. The grading plan must show all finish grades, spot elevations as necessary and existing, and new contours within the developed landscape area.

(5) Landscape Architect: A person who holds a certificate to practice landscape architecture in the state of California under the authority of the California State Board of Landscape Architects.

(6) Low-water-use-plants: Plants on MMWD low-water-use plant list, or any other plant approved by MMWD. (Generally, a plant that once established, can survive on two irrigations per month during the summer months.)

(7) Overhead Irrigation: An irrigation method that delivers water to the landscape in a spray or stream-like manner from above-ground spray heads (includes micromisters, does not include bubblers).

(8) Overspray: Water that is delivered beyond the targeted landscape area during windless hours onto adjacent pavements, walks, structures, or other non-landscaped areas during an irrigation cycle.

(9) Runoff: Irrigation water that is not absorbed by the soil or landscape area to which it is applied and which flows onto other areas.

11.60.030 Requirements for All Services

(1) Pressure Regulation. A pressure-regulating valve shall be installed and maintained by the consumer if static service pressure exceeds 80 pounds per square inch. The pressure-regulating valve shall be located between the meter and the first point of water use, or first point of division in the pipe, and set at not more than 50 pounds per square inch when measured at the most elevated fixture in the structure served. This requirement may be waived if the consumer presents evidence satisfactory to the District that excessive pressure has been considered in the design of water-using devices and that no water will be wasted as a result of high-pressure operation.

(2) Interior Plumbing Fixtures. All plumbing installed, replaced or moved in any new or existing service must meet the following requirements:

   (A) Toilets shall use 1.6 gallons, or less, of water per flush;

   (B) Shower heads shall use 3 gallons, or less, of water per minute;

   (C) Kitchen and lavatory faucets shall use 2.5 gallons, or less, of water per minute;

(D) Non-residential services with more than one shower head or one sink (lavatory) per bathroom facility shall equip these fixtures with self-closing valves.

(3) Pool Covers. Pool covers are required for all new outdoor swimming pools.

11.60.040 Landscape Requirements for Single-Family Residences (New and Modified Landscapes).

The combined size of turf areas and swimming pools for new single family residences shall be limited to not more than 25% of the total developed landscape area. When existing landscape areas are modified by the addition of turf and/or a pool, the total combined area of the turf and pool shall be no more than 25% of the total developed landscape area.

11.60.050 Landscape Requirements for All Services Other Than Single-Family Residences (New and Modified Landscapes).

(1) Turf and Swimming Pools. The combined size of turf areas and swimming pools shall be limited to not more than 25% of the total developed landscape area in services irrigated with potable water. In landscapes irrigated with reclaimed water, the combined size of turf areas and swimming pools shall be limited to not more than 40% of the total developed landscape area. In areas designated for future reclaimed water service, the 40% turf/pool limit will be allowed only if the service will have reclaimed water available within one year of the service agreement date.

(2) High-Water-Use Plants and Features. High-water-use plants, decora- tive pools (non-swimming), fountains, and water features shall be limited to not more than 10% of the total developed landscape area.

(3) Other Plants. All other plantings shall be composed of low-water-use plant materials. The District may waive this requirement if sufficient evidence is presented that the site is not suitable for such plants.

(4) Irrigation Systems. All irrigated landscaped areas will be irrigated by an automatic irrigation system which meets these requirements:

(A) Electric controller with repeat start time and multiple program potential, set for irrigation between the hours of 6 p.m. and 11 a.m. for potable water and 10 p.m. and 7 a.m. for reclaimed water;

(B) Automatic rain shut-off unit for each controller;

(C) In areas with slopes exceeding 15%, the precipitation rate shall not exceed .85 inches per hour;

(D) Under-the-head check valves, built-in spray head check valves, or in-line check valves must be installed as needed to prevent low head drainage and puddling;

(E) Separate irrigation circuit(s) must be provided for each of the following: turf, high-water-use plants, low-water-use plants, plants on drip irrigation, planting areas with different exposures, slopes and soils with different infiltration rates;

(F) Use a point application (drip, bubbler, etc.) or subsurface irrigation system where overspray, angle of slope, soil texture, or widely spaced plants make overhead irrigation impractical due to overspray, runoff, or inefficiency. (Overhead irrigation is inefficient when less than 50% of spray pattern of any head will hit mature plants.)

(G) Overhead irrigation must meet the following additional requirements:

- **(i)** Distance between spray heads on turf shall not exceed 55% of the spray diameter;

- **(ii)** Distance between spray heads elsewhere shall not exceed 70% of the spray diameter;

- **(iii)** Spray heads must be adjusted so spray radius or special pattern is within 25% of the manufacturer's rating;

- **(iv)** Spray heads must be located so that overspray will not accumulate and flow off adjacent pavements, walkways, structures, and other non-landscaped areas during an irrigation cycle.

- **(v)** Overhead irrigation is prohibited in median strips and parking islands less than eight feet wide;

- **(vi)** Planted areas which are acutely angled or irregularly shaped and which are adjacent to hardscape surfaces shall not be irrigated by an overhead system unless they are at least 120% of the spray diameter of the irrigation heads being used.

- **(vii)** Precipitation rates for all heads within each valve circuit must be matched to within 20% of one another;

(5) Soil Preparation. For overhead irrigation, soil preparation must meet the recommendations of a soils laboratory report or otherwise meet the following minimum requirements:

(A) Areas with slope ratios greater than 3:1 must be amended as recommended by a landscape architect;

(B) Areas with slope ratios of 3:1 or less must meet the following soil preparation requirements:

- **(i)** Rip or rotary cultivate existing soil to a depth of six (6) inches;

- **(ii)** Incorporate an organic amendment at the rate of 5 cubic yards per 1000 square feet into the top six (6) inches of soil.

(6) Mulching. All exposed soil surfaces of non-turf areas within the developed landscape area must be mulched with a minimum two (2) inch deep layer of organic material.

(7) Plan Review, Certification and Site Inspections.

(A) Plan Review and Certification.

- **(i)** For all services other than single family residences, applicants shall obtain approval for landscape plans for new or modified landscapes from MMWD before construction begins.

- **(ii)** All landscape plans submitted must be certified by a landscape architect that they are in compliance. The landscape architect must also sign and return a District checklist indicating compliance with District requirements. Each page of plans must also be stamped to certify compliance.

(B) It shall be the responsibility of the owner or the owner's agent to:

- **(i)** Schedule a site inspection with a landscape architect prior to installation of the irrigation system. The inspection is to verify that the installing contractor is using District approved plans for the site and that the soil preparation requirements of this section have been met. The landscape architect must complete and submit to the District a District verification form within 10 days of this inspection;

- **(ii)** Schedule an inspection with a landscape architect within 10 days of completion of work to verify compliance with the approved landscape plans. The landscape architect must complete and submit to the District a District verification form within 10 days of this inspection.

(C) District reserves the right to perform site inspections at any time before, during, or after irrigation system and landscape installation and to require corrective measures if requirements of this ordinance are not satisfied.

11.60.060 Other Provisions.

The District will consider and may allow the substitution of well-designed conservation alternatives or innovations which may equally reduce water consumption for any of the foregoing requirements of this chapter.

Section 2. The severability provisions of sections 1.10.110 of the Marin Municipal Water District Code are applicable to this Ordinance.

PASSED AND ADOPTED THIS 28th DAY OF August, 1991, by the following vote of the Board:

AYES: Boessenecker, Cronin, Miller, Morrison, Wray

NOES: None

ABSENT: None

_____
President, Board of Directors

ATTEST:

_____
Secretary, Board of Directors

MARIN MUNICIPAL WATER DISTRICT

220 Nellen Avenue
Corte Madera, CA 94925-1169
415.924.4600
FAX 415.927.4953

LANDSCAPE ARCHITECT'S CERTIFICATION STATEMENT FOR ORDINANCE No. 326

PROJECT DATA SHEET

Please complete the following information:

DISTRICT PLAN
REVIEW NUMBER: _____

PROJECT NAME: _____

PROJECT
ADDRESS: _____
_____

ACCESSORS'S
PARCEL NUMBER: _____

WATER ENTITLEMENT
FOR IRRIGATION: _____ ACRE FEET PER YEAR.

APPLICANT

NAME: _____

ADDRESS: _____
_____

PHONE: _____ FAX: _____

PDS Page 1 of 2

An Equal Opportunity/Affirmative Action Employer

APPLICANT'S AGENT (if applicable)

NAME: ______________________________

ADDRESS: ______________________________

____

PHONE:______________ FAX:______________

LANDSCAPE ARCHITECT

NAME: ______________________________

LICENSE NUMBER &
EXPIRATION DATE: ______________

ADDRESS: ______________________________

____

PHONE:______________ FAX:______________

LANDSCAPE CONTRACTOR

NAME: ______________________________

LICENSE NUMBER &
EXPIRATION DATE: ______________

ADDRESS: ______________________________

____

PHONE:______________ FAX:______________

PDS Page 2 of 2

(Effective 10/91)

MARIN MUNICIPAL WATER DISTRICT

220 Nellen Avenue
Corte Madera, CA 94925-1169
415.924.4600
FAX 415.927.4953

LANDSCAPE ARCHITECT'S CHECKLIST & CERTIFICATION FOR ORDINANCE No. 326

LANDSCAPE PLAN REVIEW

GENERAL PROJECT INFORMATION

Please complete the following information:

DISTRICT PLAN
REVIEW NUMBER: _____

PROJECT NAME: _____

PROJECT ADDRESS: _____

_____

LANDSCAPE PLAN REVIEW INSTRUCTIONS

All landscape plans submitted to the District must be certified by a landscape architect that they are in compliance with Ordinance 326. All plans must pass this review process and receive District approval before beginning construction. This checklist is for the landscape architect's review.

Please check all the boxes and fill in the appropriate blanks below. Unchecked boxes and blanks require the landscape architect to submit a written explanation why the District should consider a variance and why that requirement cannot be met.

The District will consider and may allow the substitution of well designed conservation alternatives or innovations which may equally reduce water consumption for any of the Ordinance 326 requirements. However, all written explanations, variances, substitutions, alternatives or innovations must be reviewed and approved by the District before construction begins.

LPR 1 of 6

An Equal Opportunity/Affirmative Action Employer

Please submit the following information to the District's Environmental Resources Division for their review:

1. Completed Project Data Sheet (PDS).
2. Completed Landscape Plan Review (LPR).
3. One complete set of landscape plans.

The District will notify applicant if the plans pass or fail the review. All plans that pass the review will receive a District stamp of approval on all sepia reproducible drawings before beginning construction. The plans not passing review will be resubmitted to the District until those plans do pass review.

GENERAL PLAN REQUIREMENTS

Please show plan page number(s) and drawing date of plans.

1. [ ]   Planting plan(s)   _____

2. [ ]   Irrigation plan(s) _____

3. [ ]   Grading plan(s)   _____

IRRIGATION REQUIREMENTS:

1. [ ]   The irrigation plan is drawn at the same scale as the planting plan. The irrigation plan is separate from but in the same format as the planting plan.

2.   The irrigation plan has the following information:
   a. [ ] Irrigation system point of connection;
   b. [ ] Water service pressure at point of irrigation system connection;
   c. [ ] Water meter size;
   d. [ ] Backflow prevention device(s);
   e. [ ] Major components of the irrigation system;
   f. [ ] Total precipitation rate shown in inches per hour for each valve circuit using over-head irrigation.
   g. [ ] Total flow rate (GPM) and operating pressure (PSI) for each irrigation circuit.

3.   Pressure regulation valve(s) are shown.
   a. [ ] A pressure regulation valve(s) is indicated where water pressure exceeds 80 psi. The water service pressure is ___ psi.
   b. [ ] A pressure regulation valve was not needed because water service pressure less than 80 psi. The irrigation design pressure is ___ psi.

4. [ ] Automatic controller with repeat start times and multiple program potential are shown on plan.

5. [ ] Automatic rain shut-off unit(s) are shown on plan for each controller.

6. [ ] Precipitation rates do not exceed .85 inches per hour on slopes exceeding 15%.

7. [ ] Check valves are shown on plan.

8. Separate irrigation circuit(s) are provided for the following:
   a. [ ] Turf;
   b. [ ] High-water use plants;
   c. [ ] Low-water use plants;
   d. [ ] Plants on drip;
   e. [ ] Exposure variations;
   f. [ ] Slope variations;
   g. [ ] Soils with different infiltration rates;
   h. [ ] Different precipitation rates.

9. [ ] Point application or subsurface irrigation systems are used where overspray, angle or slope, soil texture, or widely spaced plants make overhead irrigation impractical due to overspray, runoff, or inefficiency.

10. Irrigation legend has the following information:

    a. [ ] Symbols for all irrigation equipment;
    b. [ ] General description of equipment;
    c. [ ] Manufacturer name and model number;
    d. [ ] Operating pressure;
    e. [ ] Manufacturer's rated gpm per nozzle;
    f. [ ] Minimum and maximum spray radius;
    g. [ ] Manufacturer's rated precipitation rate per nozzle.

11. Overhead irrigation satisfies the following criteria:
    a. [ ] Distance between spray heads on turf does not exceed 55% of the spray diameter;
    b. [ ] Distance between spray heads elsewhere does not exceed 70% of spray diameter;
    c. [ ] Spray heads are adjusted so spray radius or special pattern is within 25% of the manufacturer's rating;
    d. [ ] Spray heads are located so overspray will not accumulate and flow off adjacent pavements, walkways, structures and other non-landscaped areas during an irrigation cycle;

e. [ ] Median strips and parking islands less than eight feet wide have no overhead irrigation;
f. [ ] Planted areas which are acutely angled or irregularly shaped and which are adjacent to hardscape surfaces are not irrigated by an overhead system unless they are at least 120% of the spray diameter of the irrigation heads being used;
g. [ ] Precipitation rates within each overhead circuit are matched to within 20% of one another.

12. [ ] All reclaimed water piping and guidelines have been met.

GRADING REQUIREMENTS:

1. [ ] The grading plan is drawn at the same scale as the planting and irrigation plans.

2. The grading plan shows the following information:

   a. [ ] Finish grades;
   b. [ ] Spot elevations as necessary;
   c. [ ] Existing and new contours within the developed landscape area.

PLANTING REQUIREMENTS:

1. [ ] The planting plan is drawn at the same scale as the irrigation and grading plan.

2. The planting plan shows the following information:
   a. [ ] New and existing trees, shrubs, groundcovers and turf areas within the developed landscape area;
   b. [ ] Plant botanical name(s), common name(s), container size(s), spacing and quantities;
   c. [ ] Property lines, streets and street names;
   d. [ ] Driveway(s), sidewalk(s) and other hardscape features as necessary;
   e. [ ] Pool(s), fountain(s), fence(s) and retaining wall(s);
   f. [ ] Existing and proposed buildings;
   g. [ ] Square footages for the various landscape hydrozones.

3. [ ] The combined area of turf and swimming pools does not exceed 25% of the total developed landscape area.
   Note: This requirement applies only to services using **potable water** for irrigation.

4. [ ] The combined area of turf and swimming pools does not exceed 40% of the total developed landscape area.
   Note: This requirement applies only to services using **reclaimed water** for irrigation.

5. [ ] This landscape is in an area designated as a **future reclaimed water service area** and will have reclaimed water available within one year of the service agreement date.

   A reclaimed water service connection date has been confirmed by the District's Reclaimed Water Section. The planting plan meets the requirements for services using reclaimed water for irrigation. Refer to Planting requirements, Item 4.

6. [ ] High-water use plants, decorative pools, fountains and water features do not exceed 10% of the total developed landscape area.

7. [ ] All other plantings are composed of low-water use plant material.

8. Planting areas with overhead irrigation with slope ratios greater than 3:1 conform to one of the following:

   a. [ ] The recommendations of a soils laboratory report and the report submitted to District for review.
   b. [ ] The recommendations of the landscape architect. Also, these recommendations can be found in either the written specifications or on the planting plan.

9. Planting areas with overhead irrigation with slope ratios less than 3:1 conform to one of the following:

   a. [ ] The recommendations of a soils laboratory report and the report submitted to District for review.
   b. [ ] Planter soils will be ripped or rotary cultivated to a depth of 6 inches and have an organic amendment incorporated into the soil at a rate of 5 cubic yards per 1000 square feet.

10. [ ] All exposed soil surfaces of non-turf areas within the developed landscape are mulched with a minimum 2 inch layer of organic material.

CERTIFICATION STATEMENT:

I hereby certify that the planting, irrigation and grading plans are accurate and follow the Marin Municipal Water District's Water Conservation Ordinance No.326.

LANDSCAPE
ARCHITECT: _____
(Print Name)

ADDRESS: _____

PHONE: _____ FAX: _____

_____
State license stamp with signature

(Effective 10/91)

MARIN MUNICIPAL WATER DISTRICT

220 Nellen Avenue
Corte Madera, CA 94925-1169
415.924.4600
FAX 415.927.4953

LANDSCAPE ARCHITECT'S CERTIFICATION STATEMENT

ORDINANCE NO. 326

FIRST SITE INSPECTION

Please complete the following information:

DISTRICT PLAN REVIEW NUMBER: _____

PROJECT NAME: _____

PROJECT ADDRESS: _____

_____

ACCESSORS'S PARCEL NUMBER: _____

LANDSCAPE SITE INSPECTION INSTRUCTIONS

It is the responsibility of the owner or the owner's agent to schedule a site inspection with a landscape architect prior to installation of the irrigation system. This checklist is for the landscape architect's first site inspection.

Please check all the boxes and fill in the appropriate blanks below. Unchecked boxes and blanks require the landscape architect to submit a written explanation why the District should consider a variance and why that requirement cannot be met.

The District will consider and may allow the substitution of well-designed conservation alternatives or innovations which may equally reduce water consumption for any of the Ordinance 326 requirements. However, all written explanations, variances, substitutions, alternatives or innovations must be reviewed and approved by the District before making changes in the field.

An Equal Opportunity/Affirmative Action Employer

The landscape architect must complete and return this certification statement form to the District's Environmental Resources Division within 10 days of inspection.

FIRST SITE INSPECTION:

1. [ ] The installing contractor has District approved planting and irrigation plans on site.

2. [ ] Soil preparation requirements have been satisfied.

CERTIFICATION STATEMENT:

I hereby certify that the above information is accurate and that this project follows the Marin Municipal Water District's Water Conservation Ordinance No.326:

INSPECTION DATE: _____

LANDSCAPE ARCHITECT: _____
(Print Name)

ADDRESS: _____
_____

PHONE: _____ FAX: _____

_____
State license stamp with signature

FSI Page 2 of 2

(Effective 10/91)

MARIN MUNICIPAL WATER DISTRICT

220 Nellen Avenue
Corte Madera, CA 94925-1169
415.924.4600
FAX 415.927.4953

LANDSCAPE ARCHITECT'S CERTIFICATION STATEMENT FOR ORDINANCE NO. 326

SECOND SITE INSPECTION

GENERAL PROJECT INFORMATION

Please complete the following information:

DISTRICT PLAN
REVIEW NUMBER: _____

PROJECT NAME: _____

PROJECT ADDRESS: _____

_____

ACCESSORS'S PARCEL NUMBER: _____

LANDSCAPE SITE INSPECTION INSTRUCTIONS

It is the responsibility of the owner or the owner's agent to schedule a site inspection with a landscape architect within ten days of completion of work to verify compliance with the approved landscape plans. This checklist is for the landscape architect's second site inspection.

Please check all the boxes and fill in the appropriate blanks below. Unchecked boxes and blanks require the landscape architect to submit a written explanation why the District should consider a variance and why that requirement cannot be met.

The District will consider and may allow the substitution of well-designed conservation alternatives or innovations which may equally reduce water consumption for any of the Ordinance 326 requirements. However, all written explanations, variances, substitutions, alternatives or innovations must be reviewed and approved by the District before making changes in the field.

An Equal Opportunity/Affirmative Action Employer

The landscape architect must complete and submit this certification statement form to the Districts Environmental Resources Division within 10 days of inspection.

SECOND SITE INSPECTION

1. [ ] The installed landscape meets all District planting requirements. (Refer to project's Landscape Plan Review form pages 4 and 5.)

2. [ ] The installed landscape meets all District irrigation requirements. (Refer to project's Landscape Plan Review form pages 2 and 3.)

CERTIFICATION STATEMENT:

I hereby certify that the above information is accurate and that this project follows the Marin Municipal Water District's Water Conservation Ordinance No.326:

INSPECTION DATE: _____

LANDSCAPE ARCHITECT: _____
(Print name)

ADDRESS: _____

_____

PHONE:_____ FAX:_____

_____
State license stamp with signature

(Effective 10/91)

Sample Combined Checklist and Water Budget Approach:

Landscape Water Conservation Ordinance
Hayward, Calif.

Features:

- city planning department as implementing agency
- modeled after California State Ordinance AB325
- requires submittal of a water budget, irrigation schedules, and estimated water use; establishes quantitative standards for plant water requirements and irrigation efficiency
- prescribes some specific criteria for design and installation
- requires submittal of a certificate of substantial completion prepared by a landscape professional
- includes checklist, certification, and submittal forms

Implementing Agency:
City of Hayward
Planning Department
25151 Clawiter Road
Hayward, CA 94545-2731
(510) 293-5276

ARTICLE 12

WATER EFFICIENT LANDSCAPE ORDINANCE

Section	Subject Matter
10-12.01	AUTHORITY
10-12.02	PURPOSE
10-12.03	APPLICATION
10-12.04	DEFINITIONS
10-12.05	REVIEW AND APPROVAL REQUIREMENTS
10-12.06	PLANTING PLAN
10-12.07	IRRIGATION PLAN
10-12.08	LANDSCAPE WATER USE STATEMENT
10-12.09	SOILS REPORT
10-12.10	IRRIGATION SCHEDULE
10-12.11	CERTIFICATE OF SUBSTANTIAL COMPLETION
10-12.12	LANDSCAPE DESIGN STANDARDS
10-12.13	IRRIGATION DESIGN STANDARDS
10-12.14	EXCEPTIONS TO DESIGN STANDARDS
10-12.15	ADMINISTRATION AND APPEAL PROCESS

ARTICLE 12

WATER EFFICIENT LANDSCAPE ORDINANCE
(Added by Ordinance No. 92-39, adopted December 15, 1992)

SEC. 10-12.01 AUTHORITY. This article is enacted pursuant to California Government Code section 65591 et seq. and is a "water-efficient landscape ordinance" adopted by a local agency under the provisions of said article.

SEC. 10-12.02 PURPOSE. The City Council finds and declares that it is in the public interest to promote water efficient landscape design, installation, and management while ensuring that the aesthetic, functional, energy, and environmental benefits of landscaping are achieved. The purpose of the regulations set forth in this article is to establish standards for designing, installing, and maintaining water efficient landscapes in new and substantially altered or expanded existing development projects.

This article shall not preclude compliance with the landscaping performance standards contained in the Hayward Zoning Ordinance, that is, Article 1 of this chapter of the Hayward Municipal Code.

SEC. 10-12.03 APPLICATION.

This article shall apply to all new and existing development projects that contain 2,500 square feet or more of new or renovated irrigated landscaped area, except that the following projects shall be exempt from this article:

(a) Homeowner-provided landscaping for a single-family lot or for a private yard within a multi-family development;

(b) Cemeteries;

(c) Registered or City-designated historic districts, sites, and structures;

(d) Ecological restoration projects that do not require a permanent irrigation system;

(e) Landscaping that is irrigated solely with reclaimed water or well water, where an irrigation connection to the City water system is not proposed; and

(f) Public parks and recreation areas, golf courses, and school playgrounds.

APPENDIX C 127

SEC. 10-12.04 DEFINITIONS. The terms used in this article shall be defined as follows:

(a) Bubbler. An irrigation head that delivers water to the root zone by "flooding" the planted area, usually measured in gallons pursuant to minute. Bubblers exhibit a trickle, umbrella, or short stream pattern.

(b) Conversion Factor (0.62). A number that converts the Landscape Water Allowance and Estimated Landscape Water Use from acre-inches per acre per year to gallons per square foot per year. The conversion factor is calculated as follows:

(325,829 gallons/43,560 square feet)/12 inches = (0.62)
325,829 gallons = one acre foot
43,560 square feet = one acre
12 inches = one foot

(c) Drip Emitter. Drip irrigation fittings that deliver water slowly at the root zone of the plant, usually measured in gallons per hour.

(d) Ecological Restoration Project. A project where the site is intentionally altered to establish a defined, indigenous, historic ecosystem.

(e) Estimated Landscape Water Use or ELWU. The annual total amount of water estimated to be needed to keep the plants in the landscaped area healthy. It is based upon the local reference evapotranspiration rate, the size of the landscaped area, the types of plants, and the efficiency of the irrigation system, as described in section 10-12.08(b).

(f) ET_o Adjustment Factor. A factor applied to reference evapotranspiration, that adjusts for plant factors and irrigation efficiency, two major influences upon the amount of water that needs to be applied to the landscape. For the purpose of this article, the plant factor shall be 0.5 and irrigation efficiency shall be 0.625. Therefore,

ET_o Adjustment Factor = (0.5/0.625) = 0.8

(g) Evapotranspiration. The quantity of water evaporated from adjacent soil surfaces and transpired by plants during a specific time, expressed in inches per day, month, or year.

(h) <u>Extra-Drought Tolerant Plant</u>. A plant that can survive without irrigation throughout the year once established, although supplemental water may be desirable during drought periods for improved appearance and disease resistance. Plants listed in <u>Water-Conserving Plants and Landscape for the Bay Area</u> (second edition), published by East Bay Municipal Utility District, that can tolerate "no water after second year" are examples of such plants.

(i) <u>Irrigated Landscaped Area</u>. All portions of a development site to be improved with planting and irrigation. Natural open space areas shall not be included in the irrigated landscaped area.

(j) <u>Irrigation Efficiency</u>. The measurement of the amount of water beneficially used by plants divided by the amount of water applied. Irrigation efficiency is derived from measurements and estimates of irrigation system characteristics and management practices.

(k) <u>Landscape Water Allowance or LWA</u>. For design purposes, the upper limit of annual applied water for the established landscaped area as specified in section 10-12.08(a). It is based upon the local reference evapotranspiration rate, the ET_o Adjustment Factor, and the size of the landscaped area.

(l) <u>Landscape Zone</u>. Portion of the landscaped area having similar microclimate and soil conditions and plants with similar water needs that are served by one or several valves with a similar type of irrigation.

(m) <u>Non-Drought Tolerant Plant</u>. A plant that will require regular irrigation for adequate appearance, growth, and disease resistance.

(n) <u>Plant Factor</u>. A factor that, when multiplied by the reference evapotranspiration rate, estimates the amount of water used by plants.

(o) <u>Precipitation Rate</u>. The depth of water applied to a given area, usually measured in inches per hour.

(p) <u>Rain Shut-Off Device</u>. A device wired to the automatic controller that shuts off the irrigation system when it rains.

(q) <u>Reference Evapotranspiration Rate or ET_o</u>. A standard measurement of environmental parameters which affect the water use of plants. ET_o is expressed in inches per day, month, or year and is an estimate of the evapotranspiration of a large field of four- to seven-inch tall, cool-season grass that is well watered. The historic ET_o for the Hayward area is approximately 42 inches per year.

(r)   Spray Sprinkler. An irrigation head that sprays water through a nozzle.

(s)   Stream Sprinkler. An irrigation head that projects water through a gear rotor in single or multiple streams.

(t)   Turf. A surface layer of earth containing mowed grass with its roots.

(u)   Water-Conserving Plant. A plant that can generally survive with available rainfall once established, although supplemental irrigation may be needed or desirable during the spring and summer months. Examples of such plants are described in Water-Conserving Plants and Landscapes for the Bay area (second edition), published by the East Bay Municipal Utility District.

SEC. 10-12.05 REVIEW AND APPROVAL REQUIREMENTS.

(a)   Prior to issuance of a building permit for a project, or as otherwise specified in the planning approval for the project, the following landscaping documents shall be submitted for review and approval by the City Landscape Architect:

(1)   Planting Plan;

(2)   Irrigation Plan;

(3)   Landscape Water Use Statement, except for developer-installed landscaping on single-family lots; and

(4)   Soils report, if required pursuant to section 10-12.09 of this article.

If a project does not require a building permit or formal planning approval, the landscaping documents shall be submitted to the City Landscape Architect for review and approval prior to the commencement of site improvements.

(b)   Prior to issuance of a Certificate of Occupancy for a development project or upon completion of a landscaping project where a building permit is not required, the following landscaping documents shall be submitted to the City Landscape Architect for review and approval:

(1)   Irrigation Schedule; and

(2)   Certificate of Substantial Completion.

(c) The documents listed in sections 10-12.05(a) and (b) shall be prepared and signed by a landscape architect, landscape designer, or irrigation designer, except that the soils report shall be prepared by a qualified soil and plant laboratory. Additionally, the documents listed in section 10-12.05(b) may be prepared by a licensed landscape contractor.

SEC. 10-12.06 PLANTING PLAN. A detailed planting plan shall be drawn at a scale that clearly identifies the following:

(a) Location of all proposed plant materials and a legend summarizing the botanical and common names, quantity, and size of plant materials;

(b) Property lines and street names;

(c) Existing and proposed buildings, structures, retaining walls, fences, utilities, paved areas, and other site improvements;

(d) Existing trees and plant materials to be removed or retained;

(e) Where landscaped areas exceed 10 percent slope, contour lines and spot elevations as necessary for the proposed finished grade;

(f) Designation of landscape zones used to calculate the estimated landscape water use pursuant to section 10-12.08(b);

(g) Details and specifications for tree staking, soil preparation, and other applicable planting work; and

(h) Where applicable, specifications for stockpiling and reapplying site topsoil or imported topsoil.

SEC. 10-12.07 IRRIGATION PLAN. A detailed irrigation plan shall be drawn at the same scale as the planting plan and shall contain the following information:

(a) Layout of the irrigation system and a legend summarizing the type and size of all components of the irrigation system, including manufacturer name and model numbers;

(b) Static water pressure in pounds per square inch at the point of connection to the public water supply (or to a water well where applicable);

(c) Flow rate in gallons per minute and design operating pressure in pounds per square inch for each valve; also, precipitation rate in inches per hour for each valve with sprinklers; and

(d) Installation details for irrigation components.

SEC. 10-12.08 LANDSCAPE WATER USE STATEMENT. The Landscape Water Use Statement shall contain the following information:

(a) <u>Landscape Water Allowance</u>. The Landscape Water Allowance for the site shall be calculated using the following formula:

$$LWA = (42 \times 0.8 \times 0.62)LA = 20.8 \times LA, \text{ where}$$

LWA	=	Landscape Water Allowance (gallons per year)
42	=	Reference Evapotranspiration Rate (ET_o) for the Hayward area (inches per year)
0.8	=	ET_o Adjustment Factor
0.62	=	Conversion Factor (to gallons per square feet)
LA	=	Total Irrigated Landscaped Area (square feet)

(b) <u>Estimated Landscape Water Use</u>. The ELWU for the site shall be based on the planting and irrigation plans prepared for the development project.

The total ELWU for a site shall consist of summing the ELWU for all landscape zones within the irrigated landscaped area. The ELWU for each landscape zone shall be calculated using the following formula:

$$ELWU = \frac{26 \times PF \times LZ}{IE}, \text{ where}$$

ELWU	=	Estimated Landscape Water Use (gallons per year)
26	=	ET_o for Hayward area times Conversion Factor to gallons per square feet (42 inches per year $\times$ 0.62)
PF	=	Plant Factor
IE	=	Irrigation Efficiency
LZ	=	Area of Landscape Zone (square feet)

For the purpose of this article, the Plant Factor (PF) shall be the following for each type of plant material, which are based on an average density planting and average microclimate conditions:

Plant Type	Plant Factor (PF)
Fescue Turf	0.7
Non-Drought Tolerant Trees, Shrubs and Ground Cover	0.7
Water-Conserving Trees, Shrubs and Ground Cover	0.5
Extra Drought-Tolerant Trees, Shrubs and Ground Cover	0.2

For the purpose of this article, Irrigation Efficiency (IE) shall be the following for each type of irrigation:

Irrigation Type	Irrigation Efficiency (IE)
Bubblers	0.85
Drip Emitters	0.85
Stream Sprinklers in planter strips 8 feet or wider	0.75
Spray Sprinklers in planter strips 8 feet or wider	0.625
Sprinklers in planter strips less than 8 feet wide	0.4

An alternative Plant Factor or Irrigation Efficiency may be approved by the City Landscape Architect in calculating the ELWU if:

(1) The factors are based on a methodology or test data that has generally been endorsed or approved by the landscape profession; or

(2) Specific microclimate or soil conditions or landscape design elements warrant the adjustment of the factors.

APPENDIX C   133

SEC. 10-12.09 SOILS REPORT. A soils report may be required by the City Landscape Architect where irrigated landscaped areas exceed 10,000 square feet or where difficult soil or landscaping conditions exist at the project site. The soils report shall describe the depth, composition, fertility, and landscaping suitability of the soil at the project site, and shall include recommendations for soil amendment, fertilizer, and other items as needed. The planting plan shall incorporate the recommendations of the soil report into the planting specifications.

SEC. 10-12.10 IRRIGATION SCHEDULE. A monthly irrigation schedule shall be prepared that covers the initial 90-day plant establishment period and the following one year period. This irrigation schedule shall consist of a table with the following information for each valve:

(a) Plant type (for example, turf, trees, water-conserving plants, non-drought tolerant plants);

(b) Irrigation type (for example, sprinklers, drip, or bubblers);

(c) Flow rate in gallons per minute;

(d) Precipitation rate in inches per hour (for valves with sprinklers only);

(e) Run times in minutes per day; and

(f) Number of watering days per week.

The irrigation schedule shall rely on the Estimated Landscape Water Use determined in section 10-12.08(b) and monthly ET_o data for the Hayward area. The irrigation schedule may also be based on a landscape irrigation audit. The amount of water applied per valve shall be adjusted as necessary for irrigation efficiency, local rainfall, microclimate conditions, depth of root zone, soil conditions, and slope.

SEC. 10-12.11 CERTIFICATE OF SUBSTANTIAL COMPLETION. The Certificate of Substantial Completion shall indicate that:

(a) The landscaping has been installed in substantial conformance to the approved planting and irrigation plans and specifications;

(b) The automatic controller has been set according to the irrigation schedule for the plant establishment period;

(c) The irrigation system has been adjusted to maximize irrigation efficiency and minimize overspray and runoff; and

(d) A copy of the irrigation schedule has been given to the property owner.

SEC. 10-12.12 LANDSCAPE DESIGN STANDARDS.

(a) <u>Landscape Water Use</u>. The Estimated Landscape Water Use shall not exceed the Landscape Water Allowance as determined in sections 10-12.08(a) and (b).

This standard shall not apply to developer-installed front yard landscaping on single-family lots. However, when a single-family project contains three or more landscaped model homes, at least one model shall comply with the above standard, with signs and information demonstrating water-efficient landscaping methods.

(b) <u>Plant Selection</u>. Plants selected for non-turf areas shall consist of plants that are well-suited to the microclimate and soil conditions at the project site. Plants with similar water needs shall be grouped together as much as possible.

For projects located at the interface between urban areas and natural open space, water-conserving plants shall be selected that will blend in with the native vegetation and are fire resistant or fire retardant. Plants with low fuel volume or high moisture content shall be emphasized. Plants that tend to accumulate an excessive amount of dead wood or debris shall be avoided.

Slope areas shall be landscaped with deep-rooting, water-conserving plants for erosion control and soil stabilization.

(c) <u>Turf Limitation and Type</u>. Turf shall be a variety with a water requirement less than or equal to Tall Fescue. Exceptions may be granted where turf will be added contiguous to an existing turf area.

Turf shall not be installed on slopes exceeding 15 percent, unless justified to match existing conditions or surrounding development.

Developer-installed front yard landscaping on single-family lots shall be limited to 50 percent turf.

(d) <u>Mulch</u>. After completion of all planting, all irrigated non-turf areas shall be covered with a minimum two-inch layer of wood chip or bark to retain water, inhibit weed growth, and moderate soil temperature. Non-porous material shall not be placed under the mulch.

SEC. 10-12.13 IRRIGATION DESIGN STANDARDS.

(a) All irrigation systems shall include a electric automatic controller with multiple program and multiple repeat cycle capabilities and a flexible calendar program.

(b) On slopes exceeding 25 percent or 4:1 grade, the irrigation system shall consist of drip emitters, bubblers, or sprinklers with a maximum precipitation rate of 0.85 inches per hour.

(c) Each valve shall irrigate a landscape zone with similar site, slope and soil conditions and plant materials with similar watering needs. Turf and non-turf areas shall be irrigated on separate valves. Drip emitters and sprinklers shall be placed on separate valves.

(d) Drip emitters or a bubbler shall be provided to each tree. Bubblers shall not exceed 1.5 gallons per minute per device. Bubblers for trees shall be placed on a separate valve unless specifically exempted by the City Landscape Architect due to the limited number of trees on the project site.

(e) Sprinklers shall have matched precipitation rates within each control valve circuit.

(f) Sprinklers located next to paving shall be pop-up heads. Pop-up heads shall have a minimum 4-inch height in turf areas and a minimum 6-inch height in ground cover areas.

(g) Check valves shall be required where elevation differences will cause low-head drainage. Pressure compensating valves and sprinklers shall be required where a significant variation in water pressure will occur within the irrigation system due to elevation differences.

(h) Spacing of sprinklers shall not exceed 1.0 times the radius of the head for square spacing and 1.2 times the radius of the head for triangular spacing. The irrigation system shall be designed for minimum run-off and overspray onto non-irrigated areas.

(i) A rain shut-off device shall be installed to prevent irrigation during rainy weather.

(j) A pressure regulator shall be provided when the static water pressure exceeds the maximum recommended operating pressure of the irrigation system.

(k) Drip irrigation lines shall be undergrounded, except for emitters and where approved as a temporary installation. Filters and end flush valves shall be provided as necessary.

(l) Valves with spray or stream sprinklers shall be scheduled to operate between 9 p.m. and 8 a.m. to reduce water loss from wind and evaporation.

(m) Program valves for multiple repeat cycles where necessary to reduce runoff, particularly on slopes and soils with slow infiltration rates.

SEC. 10-12.14 EXCEPTIONS TO DESIGN STANDARDS. Exceptions to the landscaping and irrigation standards contained in sections 10-12.12 and 10-12.13 may be granted by the City Landscape Architect where:

(a) Unique soil, site conditions, or design constraints render compliance with certain standards infeasible;

(b) The functional or recreational purpose of the landscaping warrants exceptions to specific standards; or

(c) Alternative water-efficient design techniques or materials are proposed to justify exceptions to specific standards.

SEC. 10-12.15 ADMINISTRATION AND APPEAL PROCESS. The City Landscape Architect or designee shall have the duty and authority to administer and enforce this article.

The City Landscape Architect's action to approve, conditionally approve, or disapprove landscaping documents required under this article may be appealed to the Board of Adjustments by the property owner or applicant by filing a written request with the Planning Department within 15 days of the date of written notification of said action. The matter shall be scheduled before the Board of Adjustments in a timely manner. The Board of Adjustments may approve, modify, or reverse the action of the City Landscape Architect. The action of the Board of Adjustments shall be final.

City of HAYWARD

LANDSCAPE DESIGN CHECKLIST
✔

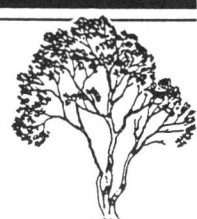

 Purpose of Checklist?

This checklist is provided to assist landscape architects and designers in preparing planting and irrigation plans that will comply with the City's landscaping standards, guidelines, and submittal requirements. The checklist is derived from the City's Zoning, Water Efficient Landscape, Tree Preservation, and Security Ordinances, Traffic Code, Design Review Guidelines, and Landscape Beautification Plan. Certain items may not pertain to your project. Please contact the City Landscape Architect, (510) 293-5215, for additional information.

 Who can prepare landscaping plans?

Landscaping plans shall be prepared by a landscape architect, landscape designer, or irrigation designer with the expertise to prepare planting and irrigation plans that comply with water efficient landscape design principles. For larger projects, the City may *require* that landscaping plans be prepared by a landscape architect. Plans shall include the signature and license or certification number, if applicable, of the design professional who prepared the plans.

 When are landscaping plans submitted?

If planning approval is required for a project (i.e., site plan review, use permit, or planned development), a *conceptual* planting plan is usually required with the development plans submitted to the Planning Department. The conceptual planting plan should indicate: general plant sizes and quantities; plant massing to comply with zoning standards, water conservation standards, and design guidelines; trees to be preserved or removed; and a suggested plant palette.

Following planning approval, *detailed* planting and irrigation plans and a Landscape Water Use Statement are to be submitted to the Building Division along with plans for a building permit, unless otherwise specified in the planning approval. Issuance of a building permit is contingent on approval of landscaping documents by the City Landscape Architect.

 What is required at completion of landscaping?

An irrigation schedule and Certificate of Substantial Completion, which shall be prepared by the design professional or licensed landscape contractor, must be submitted to the City Landscape Architect. A landscape inspection and submittal of the above documents are required prior to issuance of a Certificate of Occupancy.

Planning Department ■ 510/293-5215 ■ 25151 Clawiter Rd. ■ Hayward, CA 94545

Submittal Requirements

Planting Plan -

- [ ] Location of all proposed plant materials.
- [ ] Legend summarizing botanical and common name, quantity, and size of all plant materials.
- [ ] Property lines and street names.
- [ ] Existing and proposed buildings, structures, retaining walls, fences, utilities, paved areas, and other site improvements.
- [ ] Existing trees and plant materials to be removed or retained.
- [ ] Where landscaped areas exceed 10 percent slope, contour lines and/or spot elevations as necessary for the proposed finished grade.
- [ ] Designation of landscape zones used to calculate the Estimated Landscape Water Use (ELWU).
- [ ] Details and specifications for tree staking, soil preparation, and other planting work. (Refer to recommended standard for street tree staking.)
- [ ] Where applicable, specifications for stockpiling and reapplying site topsoil and/or imported topsoil.

Irrigation Plan -

- [ ] Layout of the irrigation system, (i.e. water meter, backflow prevention device, pressure regulator, automatic controller, main and lateral lines, valves, sprinklers, bubblers, drip emitters, quick couplers, and filters where applicable.)
- [ ] Legend summarizing the manufacturer name, model number, and size of all components of the irrigation system.
- [ ] Static water pressure (psi) at the point of connection. (Water pressure at City main available from Utilities Administration, 293-5134.)
- [ ] Flow rate (gallons per minute) and design operating pressure (psi) for each valve; also, precipitation rate (inches per hour) for each valve with sprinklers.
- [ ] Installation details for irrigation components.

- [ ] **Soils Report** (if required by City Landscape Architect) - Report shall be prepared by a qualified soil and plant laboratory. Recommendations for soil amendment and fertilizers shall be indicated on planting plan.
- [ ] **Landscape Water Use Statement** - See Attachment A
- [ ] **Irrigation Schedule** - See Attachment B. Submit Schedule when landscaping is completed, prior to issuance of Certificate of Occupancy.
- [ ] **Certificate of Substantial Completion** - See Attachment C. Submit Certificate when landscaping is completed, prior to issuance of a Certificate of Occupancy.

Development Standards

- **Setbacks** - required front, side street, side, and rear yards fully landscaped except for permitted paved areas and other approved encroachments.

  *Comment: Confirm with property owner/applicant or Planning Department regarding required setbacks for development. The **sole** use of rock or wood bark in landscaped areas is not permitted.*

- **Street Trees** - minimum one tree for every 20 to 40 lineal feet of street frontage, depending on tree species and as directed by City Landscape Architect.

  *Comment: Unless otherwise allowed, street trees shall be 24" box containers. Refer to City's List of Recommended Street Trees. City Landscape Architect may also specify a tree for certain streets:*_____

- **Parking Lot Trees** - minimum one tree for every eight parking stalls; tree wells minimum 20 square feet in area.

  Comment: Unless otherwise allowed, trees shall be minimum 15-gallon containers and located within the parking area.

- **Parking Lot Screening** - parking areas screened from street with low shrubs, walls, and/or earth berms. Maximum height of earth berms shall be 30" per City's Security Ordinance.

  Comment: Shrubs should be minimum 5-gallon containers and provide a continuous 30" high screen within two years.

- **Parcels Abutting BART Tracks** (or within 500 feet and in direct view of BART tracks) - landscape screen provided along property line.

  Comment: Screen should consist of a minimum 10-foot wide landscape strip with minimum 15-gallon evergreen trees planted 20 feet apart.

- **Curbs** - landscape areas adjoining driveways and/or parking areas separated by 6" high Class "B" Portland Cement concrete curb.

- **Drive-in Establishments** (e.g., service stations, car washes, fast-food restaurants, etc.) - check with Planning Department for specific landscaping standards.

- **Security** - landscaping will not obstruct building or parking lot light fixtures, address signs, building entrances, and windows.

- **Sight Distance** - for corner lots, within the area described below, shrubs kept to maximum 3 feet high and tree branches kept to minimum 8 feet above the grade at the center of the intersection.

Other Landscaping Requirements (e.g. conditions of approval for planning permit):

- _____
- _____
- _____

Design Guidelines

- [ ] Outdoor spaces, pathways, and edges defined with landscaping.
- [ ] Adjacent land uses buffered with landscaping.
- [ ] Landscaping complements adjacent landscaping.
- [ ] Landscaping complements architectural style and form of building, accentuates building features and entrances, and is compatible with building colors and materials.
- [ ] Parking, loading and service areas, utilities, solid building surfaces, retaining and masonry walls, and fences are screened with landscaping.
- [ ] Plants preserve required vehicular and pedestrian clearances, 13'-6" for trucks and 8'-6" for pedestrians.
- [ ] Mature plants will fit space and will not cause damage to pavement or underground utilities.
- [ ] Plants preserve sight distance at site entries/exits and internal circulation routes.
- [ ] Deep-rooted plants on slopes for erosion control; jute mesh netting or a comparable erosion control material on slopes 2:1 or steeper.
- [ ] Plants display variations in texture and form, with attention to flowering shrubs and seasonal color.
- [ ] Structures located on visually prominent sites, such as ridgelines and hill faces, are buffered with natural-looking landscaping.
- [ ] Landscaping at urban/wildland interface is fire-resistant (i.e. high moisture content or low fuel volume), conforms to fire safety standards and guidelines, and blends with natural surroundings.
- [ ] For projects located along arterial streets, street frontage landscaping is consistent with guidelines in *Landscape Beautification Plan* (LBP).

    Comment: Arterials covered by the LBP consist of Jackson Street, "A" Street, Foothill Boulevard, Hesperian Boulevard, Mission Boulevard, Winton Avenue, Mission Boulevard, Harder Road, Tennyson Road, Industrial Boulevard/Parkway, "B" Street, Second Street, Fairview Avenue, and Hayward Boulevard. Obtain a copy of the guidelines from the Planning Department.

Other Site-Specific Landscaping Considerations:

- [ ] _____
- [ ] _____
- [ ] _____

Water Conservation Standards

- Estimated Landscape Water Use (ELWU) does not exceed Landscape Water Allowance (LWA). *See Attachment A.*

- Plants well-suited to microclimate and soil conditions at site, require minimal water once established, are relatively free from pests and diseases, and are generally easy to maintain.

    Comment: Refer to EBMUD's **Water-Conserving Plants and Landscapes for the Bay Area** *or Bob Perry's* **Trees and Shrubs for Dry California Landscapes** *for recommended water-conserving plants.*

- Plants with similar water needs grouped together.

- Where turf proposed, Tall Fescue or variety with similar water requirement specified.

- Turf not proposed on slopes exceeding 15 percent.

- For developer-installed frontyard landscaping on single-family lots, turf limited to 50 percent turf.

- Pre-emergent herbicide and minimum two-inches of wood mulch specified on plans.

Irrigation:

- Automatic controller provided with multiple program and repeat cycle capabilities and a flexible calendar program.

- On slopes over 25 percent, or 4:1 grade irrigation system consists of drip emitters, bubblers or sprinklers with maximum precipitation rate of 0.85 inches per hour.

- Each valve irrigates area with similar site, slope, and soil conditions and plants with similar watering needs.

- Turf and non-turf areas irrigated on separate valves.

- Drip emitters and sprinklers on separate valves.

- Drip emitters or a bubbler provided to each tree; bubblers maximum 1.5 gallons per minute. Bubblers for trees placed on separate valve, unless otherwise permitted by City Landscape Architect.

- Sprinklers have matched precipitation rate within each valve.

- Pop-up sprinklers specified next to paving - 4" in turf, 6" in ground cover areas.

- Check valves specified where low-head drainage will occur due to elevation differences.

- Pressure compensating valves and sprinklers specified where significant variation in water pressure will occur.

- Sprinklers spaced at maximum 1.0 times radius of head for square spacing and maximum 1.2 times radius of head for triangular spacing.

- Rain shut-off device specified.

- Pressure regulator provided where static water pressure exceeds maximum recommended operating pressure.

- Drip irrigation lines undergrounded, except for temporary installations.

Tree Preservation

☐   Trees to be preserved or removed indicated on plans.

Comment: Indicate location, trunk diameter, species, and approximate dripline of trees. Retain significant trees and native vegetation that are in good condition, and avoid grading and paving within the dripline of the trees. An arborist report may be required by the City Landscape Architect.

☐   Tree protection measures noted on grading, site, and landscaping plans, if applicable.

Comment: See below for recommended minimum tree protection measures.

☐   Tree removal permit obtained prior to removing any tree 30" or larger in trunk circumference (or approximately 10" or larger in trunk diameter,) measured two feet above the ground.

Comment: Replacement trees are typically required for trees authorized for removal, which will be specified by City Landscape Architect based on condition, size, species, and location of tree(s) to be removed. Show required replacement trees on planting plan.

TREE PROTECTION NOTES

1. Tree branches that will interfere with construction equipment shall be properly pruned *prior* to beginning construction. Pruning shall be kept to a minimum and comply with accepted horticulture practices.

2. A protective fence shall be placed at the dripline of the existing trees during the construction period.

3. Soil compaction and grading shall be avoided under the dripline of the trees. Maintain positive drainage away from tree.

4. No storage of materials of equipment shall occur within 25 feet of the dripline of the trees.

5. All roots 1" or larger that must be severed shall be cut manually to produce a clean cut and treated with a tree sealant.

6. Contractor shall be responsible for providing comparable replacement trees for any existing trees that are found by the City to be irreparably damaged due to construction activity.

APPENDIX C 143

ATTACHMENT A
LANDSCAPE WATER USE STATEMENT

General Instructions:

This statement shall be submitted with the planting and irrigation plans and is the basis for achieving a water efficient landscape design. Part One should generally be completed before preparing the planting plan. Part Two should be completed after preparing a preliminary planting plan. The Landscape Water Allowance (LWA) calculated in Part One shall not exceed the Estimated Landscape Water Use (ELWU) calculated in Part Two.

For design purposes, the LWA establishes an "annual water budget" for the landscaped area within a project. It is based on evapotranspiration data (ET_o) for the Hayward area and the total square footage of irrigated landscaped area.

The ELWU is determined from the planting and irrigation plans for a project and provides an estimate of the water annually needed to keep the landscaping healthy and attractive.

A sample Landscape Water Use Statement for a hypothetical project is attached for illustration.

Preparing landscaping plans that do not exceed the LWA or "annual water budget" requires an emphasis on water-conserving plants, although a modest amount of turf or other non-drought tolerant plants will still be possible. Following are suggestions for modifying the planting and irrigation plans to reduce the landscape water use for a project, if found to be necessary:

- ❏ Group plants with similar water needs, thereby allowing for a more efficient irrigation design.

- ❏ Reduce the amount of turf or other non-drought tolerant plants. Concentrate these plants in highly visible areas or areas targeted for pedestrian or recreational activities.

- ❏ On less visible and more remote areas of a site, specify extra-drought tolerant plants that can survive with minimal water after two years. Refer to EBMUD's **Water Conserving Plants and Landscapes for the Bay Area** for suggestions.

- ❏ Where appropriate, change spray sprinklers to stream sprinklers, bubblers, or drip emitters to improve irrigation efficiency.

- ❏ In narrow planter strips (less than 8 feet wide), use drip or bubbler irrigation and avoid specifying turf.

Specific Instructions:

Part ONE

Box A - Enter the total square footage of irrigated landscaped area within the project.

Box B - Calculate the Landscape Water Allowance (LWA) for a project by multiplying the number in "Box A" by 20.8.

Part TWO

First, designate "landcape zones" on the preliminary planting plan. Each landscape zone should consist of plants with similar water needs, areas with similar microclimate (i.e., slope, exposure, wind, etc.) and soil conditions, and areas that will be similarly irrigated. A landscape zone can consist of an area served by one or several valves.

Next, complete the table in Part Two as follows:

Landscape Zone - Enter symbol corresponding to the designation on the planting plan.

Area (LZ) - Enter square footage of the landscape zone.

Plant Factor (PF) - Enter the PF from Table A below that most closely describes the type of plants in the landscape zone.

Irrigation Efficiency (IE) - Enter the IE from Table B below that describes the predominate type of irrigation in the landscape zone.

ELWU - Calculate the Estimated Landscape Water Use (gallons per year) for each landscape zone using the following formula:

$$ELWU = \frac{LZ \times PF \times 26}{IE}$$

Totals -
a) Total the square footage of all landscape zones, which should equal the total irrigated landscaped area shown in Part One, Box A.

b) Total the ELWU for all landscape zones, which shall not exceed the LWA shown in Part One, Box B.

TABLE A - Plant Factors	
Plant Type	*PF*
Fescue Turf	0.7
Non-Drought Tolerant Plants	0.7
Water-Conserving Plants	0.5
Extra Drought Tolerant Plants	0.2

TABLE B - Irrigation Efficiency	
Irrigation Type	*IE*
Bubblers	0.85
Drip Emitters	0.85
Stream Sprinklers (in planter strips 8 feet or wider)	0.75
Spray Sprinklers (in planter strips 8 feet or wider)	0.625
Sprinklers (in planter strips less than 8 feet wide)	0.4

APPENDIX C 145

EXAMPLE

City of HAYWARD

*LANDSCAPE WATER USE
STATEMENT*

Project Name: Fashion Elite Commercial Building
Project Address: 21215 Main Street
Hayward, CA 94541
Prepared by:

Creative Landscape Designs	CLA #1956
Name	*License or Cert. No. (if applicable)*
195 Garden Lane	(510) 786-5678
Address	*Telephone Number*
Hayward, CA 94541	July 15, 1992
	Date

PART ONE — *Landscape Water Allowance*

Total Irrigated Landscaped Area
(square feet)

Box A: 8,873

X 20.8

Landscape Water Allowance
(Gallons per Year)

Box B: 184,558

PART TWO — *Estimated Landscape Water Use*

$$\text{ELWU} = \frac{LZ \times PF \times 26}{IE}$$

Landscape Zone	Area (LZ) (square feet)	Plant Factor (PF)	Irrigation Efficiency (IE)	ELWU* (Gallons/Year)
A	3,113	0.2	0.85	19,044
B	1,943	0.5	0.85	29,716
C	2,592	0.5	0.75	44,928
D	1,112	0.7	0.625	32,381
E	113	0.7	0.625	3,291
Total	8,873			129,360

146 WATER-EFFICIENT LANDSCAPE GUIDELINES

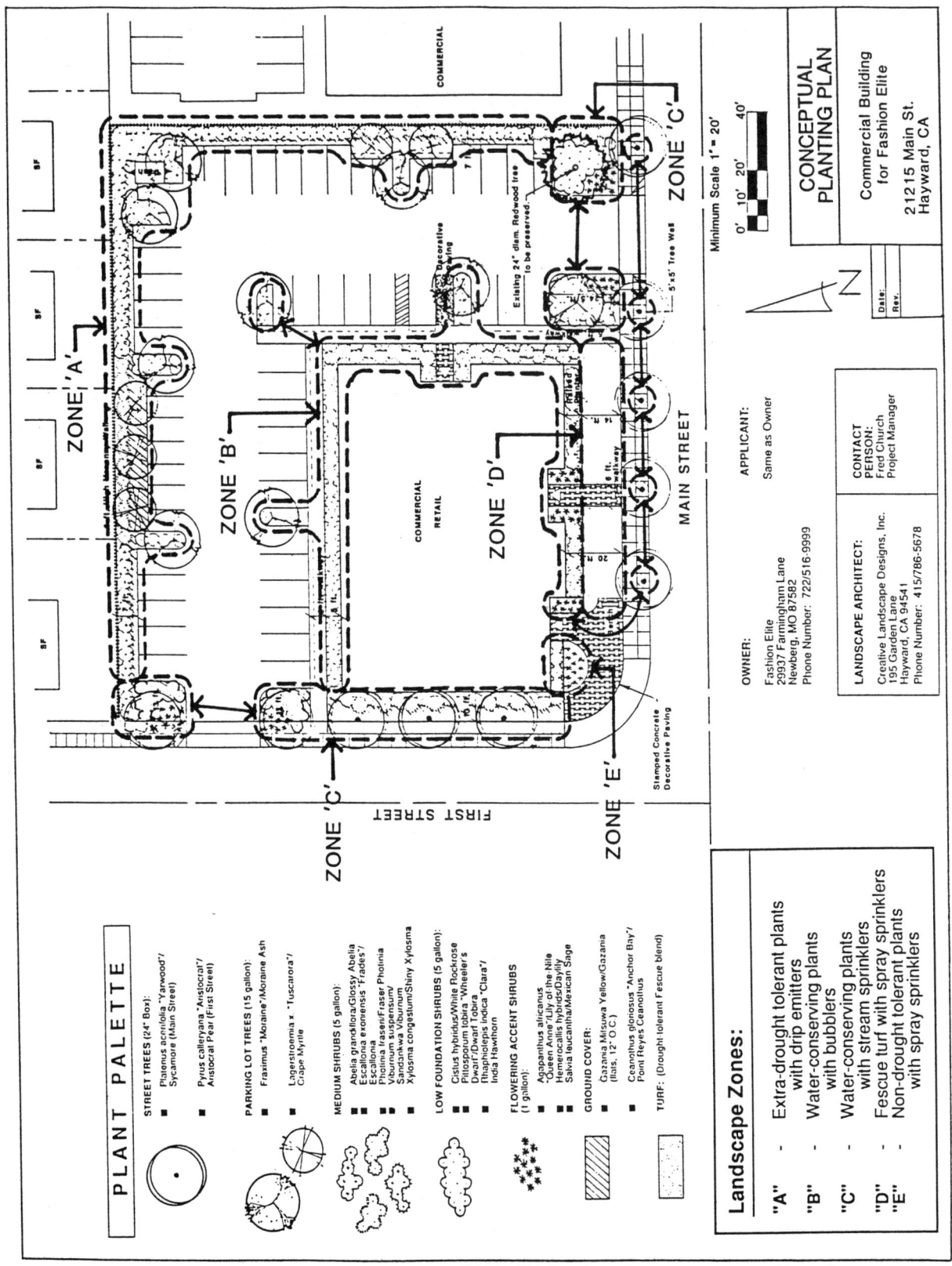

EXAMPLE: Landscape Water Use Statement

City of HAYWARD

LANDSCAPE WATER USE STATEMENT

Project Name: _____
Project Address: _____
_____

Prepared by:

_____	_____
Name	*License or Cert. No. (if applicable)*
_____	_____
Address	*Telephone Number*
_____	_____
	Date

PART ONE   *Landscape Water Allowance*

Total Irrigated Landscaped Area (square feet)   Box A [ ]

X   20.8

Landscape Water Allowance (Gallons per Year)   Box B [ ]

PART TWO   *Estimated Landscape Water Use*

$$\text{*ELWU} = \frac{LZ \times PF \times 26}{IE}$$

Landscape Zone	Area (LZ) (square feet)	Plant Factor (PF)	Irrigation Efficiency (IE)	ELWU* (Gallons/Year)
Total				

ATTACHMENT B
IRRIGATION SCHEDULE

General Instructions:

A monthly irrigation schedule shall be prepared to cover the initial 90-day plant establishment period and the following one-year period. The irrigation schedule shall be prepared by a landscape architect or designer, an irrigation designer, or a licensed landscape contractor. Attached is a suggested form for the irrigation schedule. The preparer may use this form or follow another appropriate format.

The irrigation schedule shall rely on the Estimated Landscape Water Use (ELWU) that was calculated for the project during the preparation of the landscaping plans. The schedule should also rely on monthly reference evapotranspiration (ET_o) data for the Hayward area, which is provided below. Once established, Tall Fescue turf can be maintained in an attractive manner at approximately 70 percent of the ET_o rate under normal weather conditions. Water-conserving plants typically need 50 percent or less of the ET_o under normal weather conditions. The amount of water applied for each valve should also be adjusted for irrigation efficiency, local rainfall, specific site conditions (e.g., exposure, slope, etc.) depths of root zone, and soil conditions (e.g., water holding capacity, and infiltration rate). Ultimately, the amount and frequency of irrigation will need to be monitored regularly to adjust for plant growth, climatic changes, and site conditions.

For valves with overhead spray or stream sprinklers, set valves to operate between 9 p.m. and 8 a.m. to reduce water loss from wind and evaporation. Early morning irrigation is recommended for turf and ground cover. On slopes and soils with slow infiltration rates, program valves for multiple repeat cycles to reduce run-off.

Estimated Monthly ET_o for Hayward Area* (inches per year)												
Jan	Feb	Mar	Apr	May	Jun	Jul	Aug	Sep	Oct	Nov	Dec	Ann. ET_o
1.5	1.5	2.8	3.9	5.1	5.3	6.0	5.5	4.8	3.1	1.4	0.9	41.8

* Based on historical data, extrapolated from 12-month normal year ET_o maps and U.C. publication 21426.

NOTE: The City of Hayward is presently under an emergency water rationing program. Landscape water use is limited to 1" per square foot per week for turf areas and 1/2" per square foot per week for non-turf areas. Excess water use charges are levied for exceeding water allotments. Contact Utilities Administration, 293-5134, for more information.

Specific Instructions:

A. **Valve or Station Number** - Shall correspond to irrigation plan.

B. **Plant Type** - *Indicate either:*

    T - Trees Only

    WC - Water-conserving trees, shrubs, and/or groundcover

    ND - Non-drought tolerant trees, shrubs, and/or groundcover

    GC - Ground cover only

    L - Turf

C. **Irrigation Type** - *Indicate either:*

    SP - Spray Sprinklers

    ST - Stream Sprinkler

    B - Bubblers

    D - Drip Emitters

D. **Flow Rate** - Indicate total gallons per minute or hour flowing through valve during normal operation (available on irrigation plan).

E. **Precipitation Rate** - For valves with spray or stream sprinklers *only,* indicate the average precipitation rate in inches per hours (available on irrigation plan, from irrigation manufacturer, or through field test.)

F. **Month** - Begin irrigation schedule with the month that landscaping work is completed.

G. **Run Time** - Indicate total minutes per day valve will be operating.

H. **Number of Day/Week** - Indicate number of days per week valve will be scheduled to operate.

150 WATER-EFFICIENT LANDSCAPE GUIDELINES

City of HAYWARD
MONTHLY IRRIGATION SCHEDULE

Project Name: _____

Project Address: _____

Prepared by: _____
 Name

_____ *License or Certification No. (if applicable)*
 Address

_____ *Telephone Number*

_____ *Date Prepared*

Valve or Station Number (A)	Plant Type (B)	Irrigation Type (C)	Flow Rate (D)	Precipitation Rate (E)	Initial Plant Establishment Period (3 Mos.)	Following One-Year Period (12 Months)
1						
2						
3						
4						
5						
6						

F — Month
G — Run Time (Minutes per Day)
H — Days per Week

NOTE: This irrigation Schedule should be used as a guide. The landscaping should be monitored regularly and the schedule adjusted as needed for plant growth, local rainfall, and climatic conditions. Check irrigation system frequently to minimize run-off and overspray. Schedule valves with sprinklers to irrigate between 9 PM and 8 AM to reduce water loss from wind and evaporation.

City of HAYWARD

ATTACHMENT C
CERTIFICATE OF SUBSTANTIAL COMPLETION

Project Name: _____

Project Address: _____

Building Permit #: _____ **Planning Permit #:** _____
(If applicable)

I/We hereby certify the following:

1. The landscaping work for the above project has been completed in substantial conformance to the City-approved planting and irrigation plans and specifications;

2. The automatic controller has been set according to the approved irrigation schedule for the plant establishment period;

3. The irrigation system has been adjusted to maximize irrigation efficiency and minimize overspray and runoff; and

4. A copy of the irrigation schedule has been given to the property owner.

Comments: _____
_____
_____

This Certification prepared by: (check whichever applies)

  ❑ Landscape Architect   ❑ Irrigation Designer or Consultant

  ❑ Landscape Designer   ❑ Licensed Landscape Contractor

Signature: _____ **Date:** _____

Address: _____
     _____ **License or Certification No.**
              (If applicable)

Telephone: _____

Sample Water Budget Approach:

East Bay Municipal Utility District
Irrigation Incentive Program

Features:

- public water utility as the implementing agency
- establishes water allocations for irrigation accounts using the simple water budget formula
- requires submittal of documented irrigated area
- uses water-conserving rate structure as incentive
- includes a submittal form

Implementing Agency:
East Bay Municipal Utility District
Office of Water Conservation, MS107
375 11th St.
Oakland, CA 94607-4240
(510) 287-0590

IRRIGATION INCENTIVE PROGRAM

FACT SHEET

WHAT IS IT?

The Irrigation Incentive Program was initiated in May 1991 and is designed to encourage long-term efficient irrigation for new and existing irrigation customers and to recognize existing customers who actively conserved in 1986, the basis year for EBMUD's drought program.

HOW DOES IT WORK?

Program participants receive water allocations based upon the inland or coastal Evapotranspiration (ET) rates of a cool-season lawn. Simply put, ET is the amount of water actually <u>used</u> by the plant and the amount lost through the leaves and soil through evaporation. Participants provide EBMUD with the landscape area by irrigation meter in square feet. The District then allots the amount of water necessary to keep the irrigated area healthy.

WHY BE INVOLVED?

The advantages are twofold: since allocations during the current drought period are based upon a 30% cutback over 1986 use levels, those who successfully conserved in 1986 would be unfairly burdened with a lower water allocation compared to someone who made no effort in 1986. Participation in the Incentive Program recognizes past efficiencies and allows 100% of ET, or the amount of water the landscape actually needs. Those who have no 1986 history or those who simply want to make a greater effort at conservation, have a practical guide to efficiently maintain the growth and vigor of their landscape. Since this program provides irrigation efficiency standards, the landscape manager can determine how efficiently water is being applied to the landscape and make changes if necessary.

WHO SHOULD BE INVOLVED?

The program is currently open to EBMUD "Irrigation Only" accounts. All new EBMUD irrigation accounts should participate since the alternative is a much lower allocation: 30% less than ET instead of 100% of ET for accounts with no prior history. Those who have 1986 use histories should carefully research to determine that the Incentive allocations are their best alternative.

IS IT FINANCIALLY VIABLE FOR THE CUSTOMER?

Yes. The program by its nature encourages efficient irrigation and the drought rates are the lowest possible with the current rate structure. It does require landscape management skills, however, to achieve optimum control of irrigation scheduling in order to fully realize financial benefit on the Incentive Program.

WHO TO CONTACT?

Direct any questions to:

> EBMUD
> John Passama or Andrea Pook
> P.O. Box 937
> Alamo, CA 94507
> Ph# (510)820-2760

APPENDIX C   155

INSTRUCTIONS FOR INCENTIVE PROGRAM SUBMITTAL

A survey form and documentation of irrigated acreage must be submitted for participation in the Incentive Program. The following information is provided to assist you in preparing a submittal. You may contact Andrea Pook or John Passama at the EBMUD Water Conservation Office (510)820-2760 for assistance.

Step 1. **LOCATE YOUR METER**

Most accounts have one meter. The meter address for each account is included in this packet to assist you in locating your meter.

If you have more than one account and, therefore, more than one meter, please fill out one form for each account.

Step 2. **CALCULATE IRRIGATED AREAS**

The irrigated area includes turf and other landscaped areas (trees, shrubs, groundcovers, and annuals).

To determine the area per meter, you must trace the irrigation valves back to their corresponding meter. This can be done in the field or with the aid of irrigation plans. Simply draw a line on your plan outlining the area irrigated by each meter and measure in square feet.

Three methods can be used to measure these areas. Note that all measurements should be in square feet.

a. **From Landscape or Irrigation Plans:**

Use only plans drawn to standard architectural and engineering scales. The scale must be indicated by notation and a scale bar. Areas can be measured using an architect's or engineer's scale, or for irregular areas, a planimeter. Any changes to the landscape that affect these measurements must be taken into account.

b. **Aerial Photo(s):**

Aerial photo prints can be purchased from the Engineering or Planning Department of your city or from a local aerial photo company. Measure the irrigated area by scale or planimeter. Aerial photo suppliers also offer measurement services.

INSTRUCTIONS FOR INCENTIVE PROGRAM SUBMITTAL

c. On-site Survey:

In the event that plans are not available, landscape areas can be measured in the field by qualified personnel. A diagramatic plan documenting the survey must be submitted with the survey form.

Step 3. SUBMIT SURVEY FORM AND DOCUMENTATION

Complete the survey form and send with documentation to:

> EBMUD
> Irrigation Incentive Program
> P.O. Box 937
> Alamo, CA 94507

Plans or photos must include the following:

1. Meter location(s) with corresponding account number(s).

2. A delineation of the area served by this meter with the related square footage.

3. The scale must be written and indicated by a bar scale.

Please fold to 8 1/2" x 11" size for submittal.

AP/MH;nt

IRRIGATION INCENTIVE PROGRAM
SURVEY FORM

Customer Name:_____

Customer Address:_____

_____

Customer Contact:_____ Phone #:_____

Site Address:_____

_____

Account Number:_____

Total irrigated area (all planting and turf) in square feet served by this account:

_____

What method was used to determine landscaped areas?

_____ Measurements taken from landscape or irrigation plan.

_____ Measurements taken from aerial photo.

_____ On-site survey.

Additional Comments:

_____

_____

_____

_____

_____

Was this plan prepared by someone other than yourself?

_____ Yes   _____ No

If you checked yes, preparer must also sign the certification on the reverse side.

One copy of the landscape irrigation plan, aerial photo or on-site survey must be included with this form for EBMUD review and verification.

CERTIFICATION

I certify that the information entered on this form is accurate to the best of my knowledge.

_____
Customer Name (print)

_____   _____
Customer Signature   Date

If prepared by a third party:

_____
Name of Preparer (print)

_____
Title

_____   _____
Preparer Signature   Date

Return to:

EBMUD
Irrigation Incentive Program
P.O. Box 937
Alamo, CA 94507

Appendix D

State Resources for Landscape Standards Information (partial list)

Arizona:
City of Phoenix, Water Conservation
455 N. 5th Street, 3rd Floor
Phoenix, AZ 85004
(602) 261-8369

California:
Office of Water Conservation
California Department of Water Resources
P.O. Box 942836
Sacramento, CA 94236-0001
(916) 653-7366 (O)
(916) 653-4275 (FAX)

Colorado:
Water Resource Specialist, Office of Water Conservation
State of Colorado
1313 Sherman Street, Rm. 721
Denver, CO 80203
(303) 866-3441 (O)
(303) 866-4474 (FAX)

Connecticut:
State of Connecticut
Dept. of Health Services, Water Supply Section
150 Washington Street
Hartford, CT 06106
(203) 566-1253

Florida:
Southwest Florida Water District
P.O. Box 24680
West Palm Beach, FL 33416
(407) 687-6785

Georgia:

Georgia Water Wise Council & Water Conservation Coordinator
Cobb County/Marietta Water Authority
1660 Barnes Mill Road
Marietta, GA 30062
(404) 426-8788 (O)
(404) 426-9092 (FAX)

New Mexico:

Water Resource Specialist
State Engineer Office
P.O. Box 25102
Santa Fe, NM 87504
(505) 827-3879 (O)
(505) 827-6188 (FAX)

Oregon:

City of Portland Water Department
1120 SW Fifth Avenue, Rm. 611
Portland, OR 97204
(503) 796-7457

Texas:

Extension Horticulturist
Texas A & M University
225 Horticulture/Forestry Science Bldg.
College Station, TX 77843-2134
(409) 845-7341

Washington:

Conservation Office
Seattle Water Department
710 Second Avenue, 10th Floor
Seattle, WA 98104
(206) 684-5879

Additional Sources of Information

Ball, Ken. 1990. *Xeriscape™ Programs for Water Utilities*. American Water Works Association, Denver, Colo.

Beard, J.B. 1990. Contribution of Turfgrass to Xeriscape Strategies. A speech delivered at CONSERV90, Phoenix, Ariz. (August 14, 1990.)

Blanchfield, P.G. 1992. AB325 and The Landscape Architects. A speech from the Water-Efficient Landscape Conference. Oakland, Calif. (February 1992.)

Bressan, Tom. 1992. The Use of Emitter Lines In Landscape Irrigation. *Landscape and Irrigation*, 16:3:54–56, 58.

Coate, Barrie, principle author and technical consultant. 1990. *Water Conserving Plants and Landscapes for the Bay Area*. East Bay Municipal Utility District, Oakland, Calif.

East Bay Municipal Utility District. 1991. Drip Irrigation Guidelines. Office of Water Conservation, Oakland, Calif.

Ellefson, C.L.; Stephens, T.L.; & Welsh, D. 1992. *Xeriscape Gardening: Water Conservation for the American Landscape*. Macmillan Publishing Co., New York.

Evolution of DSM: A Utility Segmentation Framework. 1992. Electric Power Research Institute Document #TR-100344, Palo Alto, Calif.

Ferguson, B.K. 1987. Water Conservation Methods In Urban Landscape Irrigation: An Exploratory Overview. Paper No. 85180 of the *Water Resources Bulletin*. (February, 1987.)

Gregory, James. 1991. An Expert's Approach To Low-Volume Irrigation. *Landscape and Irrigation*, 15:6:52–54.

Harivandi, Ali. 1991. Effluent Water for Turfgrass Irrigation. University of California Cooperative Extension Leaflet 21500, Oakland, Calif. (June 1991.)

Improving Irrigation Performance: Guidebook of Procedures and Equipment. 1992. Los Angeles Department of Water and Power, Los Angeles, Calif. (January, 1992.)

Jensen, Berman, & Allen, eds. 1990. *Evapotranspiration and Irrigation Water Requirements*. American Society of Civil Engineering, New York.

Jorgenson, G. & Solomon, K.H. 1990. Evaluating Subsurface Drip Irrigation for Turfgrass: An Interim Report. Center for Irrigation Technology, Fresno, Calif.

Kah, G.F. & Walker, R.E. 1990. *Landscape Irrigation Auditor Handbook*. Water Conservation Office, Department of Water Resources, State of California, Sacramento, Calif. (June 1990.)

Kourik, Robert. 1992. *Drip Irrigation for Every Landscape and All Climates*. Metamorphic Press, Santa Rosa, Calif.

Lewis, Randy. 1991. *Water Conserving Landscaping*. Lewis Design.

Lindsey, Patricia. 1991. Methods for Estimating Water Needs and Adequate Soil Volumes for Urban Trees. *Growing Points*. University of California Cooperative Extension, County of Marin, 28:3:1–3.

McClurg, Sue. 1992. Urban Water Costs. *Western Water*, Water Education Foundation, March/April:4–11.

Moss, L.R.; Withers, L.M.; & Cassandro, N.A. 1991. Landscape Water Conservation Handbook. Central Basin Municipal Water District and West Basin Municipal Water District. Lawrence R. Moss asnd Associates, Downey, Calif. (February 1991.)

Oliphant, J.C. 1991. Sprinkler Coverage: Getting The Big Picture Through Software. *Landscape and Irrigation*, 15:12:60.

Perry, Bob. 1981. *Trees and Shrubs for Dry California Landscapes, Plants for Water Conservation.* Land Design Publishing. San Dimas, Calif.

Selecting The Best Turfgrass. University of California Cooperative Extension, Leaflet 2589. Division of Agriculture and Natural Resources, Oakland, Calif.

Snyder, R.L. et al. 1991. Turfgrass Irrigation Scheduling. University of California Cooperative Extension, Leaflet 21492. Division of Agriculture and Natural Resources, Oakland, Calif. (July 1991.)

Sustainable Landscaping Guideline Manual. 1990. City of Irvine Community Development Department, Irvine, Calif.

Weinberg, S.S. & Roberts, J.M., eds. 1988. *Irrigation Volume Three of Handbook of Landscape Architectural Construction.* Landscape Architecture Foundation, Washington, D.C.

Glossary

aerification *See* coring.

adjusted water budget A quantity of water used to maintain a landscape based on evapotranspiration and area; adjusted to reflect an efficiency standard.

adjustment factor A decimal fraction used to modify reference evapotranspiration to reflect an efficiency standard.

application rate The rate of delivery by an irrigation circuit; in inches for sprinkler irrigation, in gallons for drip irrigation.

area Square footage or acreage measured and calculated from scale plans, photographs, or from on-site measurements.

area takeoff Calculations of area based on measurements from plans drawn to scale.

arid climate A climate characterized by less than 10 in. of annual rainfall.

as-built plans Revised site plans reflecting the actual conditions of a landscape installation.

beneficial rainfall *See* effective precipitation.

bluegrass A variety of cool-season turfgrass of the genus *Poa*. Commercially produced turfgrass mixes usually are a blend of varieties.

bubbler A type of sprinkler head that delivers a relatively large volume of water to a level area where standing water gradually infiltrates into the soil. The flow rate is large relative to the area to which the water is delivered. Bubblers are used to irrigate trees and shrubs.

building footprint The dimensions and area of a building foundation as represented in a two-dimensional plan.

catch-can test Measurement of precipitation from a sprinkler system, taken by placing graduated containers at evenly spaced intervals throughout an irrigated area.

Ccf A unit of water equal to 100 cubic feet of water or 748 gal; sometimes used as a billing unit by water purveyors.

central irrigation control A computerized system for programming irrigation controllers from a central location; uses personal computer and radio waves or hard wiring to send program information to controllers in the field.

check valve A device that prevents drainage of water from the low points of an irrigation circuit after irrigation.

checklist approach A method of establishing water-efficiency standards by prescribing specific criteria for landscape design, installation, and management.

commercial landscape A landscape adjacent to a facility used for commercial purposes, including offices and retail centers.

conversion factor A decimal fraction used to convert inches of evapotranspiration to gallons (0.623) or to convert inches of evapotranspiration to 100 cubic feet of water (0.00083).

cool-season turfgrass Turfgrass that does not ordinarily lose its color unless the average air temperature drops below 32°F (0°C) for an extended period; is not usually damaged by subfreezing temperatures. Cool-season grasses grow actively in cool weather of spring and fall and slowly in summer heat.

coring Mechanical cultivation of turfgrass using hollow tines to remove cores of turf; improves soil texture and increases air and water movement.

crop coefficient (K_c) A factor used to adjust reference evapotranspiration and calculate water requirements for a given plant species.

dedicated metering Metering of water service based on a single type of use, as in separate metering for landscape irrigation only.

developer-provided landscaping When developers contract for the design, construction, and management of residential landscapes.

distribution uniformity A measure of the efficiency of overhead irrigation calculated by analyzing the results of catch-can tests or by applying a formula to the dimensions and specifications of an irrigation plan.

drip irrigation The slow, accurate application of water to plant root zones with a system of pipe and emitters usually operated under reduced pressure.

drought An extended period of below-average precipitation resulting in increased demand and/or above-average reduction of water-storage levels.

dual and multiple programming The capacity of an irrigation controller to schedule the frequency and duration of irrigation cycles to meet varying water requirements of plants served by a system. Grouping plants and laying out irrigation circuits by similar water requirements facilitate multiple programming.

effective precipitation Rainfall that offsets evapotranspiration losses during a given time period; rainfall that enters the soil and is available to plants when needed. In a given day, effective precipitation is less than, or equal to, the daily evapotranspiration.

efficiency standard A value or criteria that establishes levels or conditions of water use in the ornamental landscape.

emitter A drip irrigation component that dispenses water to plants at a predictable rate, measured in gallons per hour.

evapotranspiration (ET) A measure of water depletion from the soil due to evaporation from the soil surface and transpiration through plant foliage.

ET *See* evapotranspiration.

factored pooling A process of assigning allocations by estimating the rates of water use and calculating allocations before consumption and billing.

flow rate The amount of water dispensed by an irrigation pipe, head, or emitter, measured in gallons per minute or gallons per hour.

flush valve A valve used to expel water and sediment from irrigation lines.

functionally required area A landscape design term used to describe a portion of an ornamental landscape intended to serve a specific function, such as pedestrian traffic, sport, or recreational activities.

gph gallons per hour.

gpm gallons per minute.

grading The design of landscape contours to accommodate a site's uses and provide adequate storm drainage; also, the process of constructing desired landscape gradients.

grading plan A plan drawn to scale that expresses the designed landscape gradient and elevation using contour lines or numeric notation of elevations.

green industry The trades, professions, and disciplines related to landscape and irrigation research, design, installation, and management.

hardscape Landscaping constructed from nonliving materials, such as concrete, brick, and lumber.

hardware efficiency A value as a percentage or a decimal fraction that represents the portion of water applied by an irrigation system that benefits the intended plants.

head radius The radius of the circular arc pattern of an overhead irrigation nozzle or sprayer.

head-to-head spacing Spacing irrigation heads so that the pattern of precipitation from one head completely overlaps the area between it and an adjacent head.

high-water-use landscape Landscapes with plants and features using water that require 50 to 80 percent of reference evapotranspiration to maintain optimum appearance.

historic basis Past water consumption used to calculate water allotments.

hydrozone A portion of a landscape area having plants with similar water needs that are either not irrigated or irrigated by a circuit or circuits with the same schedule.

hydrozoning The design practice of grouping plants by similar water requirements to maximize potential efficiency of irrigation.

impact head A type of single-stream rotor that uses a lever driven by its impact on the stream of water to rotate a nozzle in a full circle or arc. Impact heads have large radii and relatively low precipitation rates but do not provide matched precipitation rates for varying arc patterns.

industrial landscape A landscape next to an industrial facility.

infiltration rate The rate at which water permeates soil.

irrigated area The portion of an ornamental landscape that requires supplemental irrigation, usually expressed in square feet.

irrigation circuit A group of irrigation components, including heads or emitters and pipes, controlled and operated simultaneously by a remote control valve. Also, the area served by an irrigation circuit.

irrigation controller A mechanical or electronic clock that can be programmed to operate remote-control valves.

irrigation cycle A scheduled application of water by an irrigation circuit defined by a start time and its duration. Multiple cycles can be scheduled, separated by time intervals, to allow infiltration of applied water.

irrigation efficiency A value representing the amount of water beneficially applied, divided by the total water applied. Also, the product of decimal equivalents for hardware efficiency and management efficiency.

irrigation plan A two-dimensional plan drawn to scale expressing the layout of irrigation components and component specifications. Layout of pipes may be depicted diagrammatically, but location of irrigation heads and irrigation schedules should be specified.

irrigation zone *See* hydrozone.

landscape area The combination of irrigated area, nonirrigated planted area, water features, hardscape, and natural undeveloped area.

landscape coefficient A factor used to modify reference evapotranspiration and to calculate water requirements for a hydrozone.

landscape water requirements A measure of supplemental water needed to maintain the optimum health and appearance of landscape plantings and water features.

limited turf areas Restriction of turfgrass to a prescribed fraction of the landscape area.

looped pooling The process of assigning one water allocation to a group of metered accounts in a looped system. A looped system involves a set of meters in service to the same irrigation main. Total consumption for all accounts is compared to the water allocation.

low head drainage Drainage of water from irrigation lines at the low elevations in an irrigation circuit.

low-water-use plants Plants requiring less than 30 percent of reference evapotranspiration to maintain optimum appearance.

management efficiency A percentage or a decimal fraction of total water applied through irrigation that represents the portion beneficially applied through scheduling, maintenance, and repair of irrigation systems.

matched precipitation rates Equal water-delivery rate of sprinkler irrigation heads with varying arc patterns within an irrigation circuit. Matched precipitation rates are important to achieve uniform distribution of water.

mediterranean climate A climate that occurs on the western shores of the continents, characterized by moderate temperatures throughout the year, annual drought, and a rainy season.

medium-water-use plants Plants requiring 30 to 50 percent of reference evapotranspiration to maintain optimum appearance.

microclimate The climate of a specific place within a given area.

mulch A protective covering of various substances, especially organic, placed on the earth around plants to reduce weed growth and evaporation of moisture from the soil surface and to maintain even temperatures around plant roots.

multiple start times An irrigation controller's capacity to accept programming of more than one irrigation cycle for a circuit in a given day.

native and adapted plants Plants indigenous to an area or from a similar climate that require little or no supplemental irrigation once established.

native landscape Landscapes composed of native plant communities.

nonfunctional landscaping A built landscape, designed for aesthetics rather than to support a practical use or activity.

operating pressure Water pressure measured in pounds per square inch (psi) required for proper function of irrigation system components.

overspray Application of water via sprinkler irrigation to areas other than the intended area.

percent switch A feature of an irrigation controller that allows percent changes in the duration of programmed irrigation.

plan view A two-dimensional, plane surface representation of a landscape; a bird's-eye view.

plant factor A decimal fraction that represents a portion of reference evapotranspiration and a standard for plants' water requirements.

plant water requirements Depletion of moisture from the soil around plant roots due to evapotranspiration.

planting plan A two-dimensional plan drawn to scale that shows the layout of landscape plantings and plant specifications.

point-source emitter A drip-irrigation component that delivers water from a single orifice at a predictable rate, usually measured in gallons per hour.

pooling A process whereby water allocations from several different irrigation accounts are reassigned to better fit actual watering needs.

practical turf areas A landscape design and management concept promoting turf only in those areas of the landscape that are functional and the efficient management of supplemental irrigation required for those areas.

precipitation rate Application rate for sprinkler irrigation.

pressure-compensating emitter A drip-irrigation emitter designed to deliver water at a consistent flow rate under a range of operating pressure.

pressure loss The reduction in water pressure due to friction of water against the inner walls of pipe and components.

pressure reducer An irrigation system component that reduces the downstream pressure of water moving to irrigation lines.

public landscape Public street medians, parks, recreational facilities, and landscapes next to a public facility.

rain shutoff device A device connected to an irrigation controller that overrides scheduled irrigation when significant precipitation has been detected.

reclaimed water Treated, recycled wastewater.

reference evapotranspiration (ET_O) The evapotranspiration of a broad expanse of well-watered, 4-to-6-in.-tall cool-season grass.

remote-control valve An electric solenoid valve, wired to an irrigation controller, that controls the flow of water to an irrigation circuit.

runoff Surface drainage of irrigation water from the intended area.

semiarid climate A climate characterized by 10 in. to 20 in. of annual rainfall.

simple water budget The product of reference evapotranspiration, irrigated area, and a conversion factor.

site area The total area of a development site including building footprints, roadways, and parking areas, expressed in square feet or acres.

soil amendment Organic and inorganic materials added to soils to improve texture, nutrients, moisture holding capacity, and infiltration rates.

soil improvement The addition of soil amendments to planted areas.

soils report A report by a soils engineer indicating soil type(s), soil depth, uniformity, infiltration rates, and pH for a given site.

spray head A sprinkler irrigation nozzle installed on a riser that delivers water in a fixed pattern. Flow rates of spray heads are high relative to the area covered by the spray pattern.

spray irrigation Sprinkler irrigation using spray heads on fixed or pop-up risers and having relatively high precipitation rates.

sprinkler irrigation Overhead delivery of water using bubblers, spray heads, stream rotors, or impact heads. Precipitation rates will vary depending on system layout and type of head used.

storm drainage Surface movement of water due to storms.

stream rotors Sprinkler irrigation heads that deliver rotating streams of water in arcs or full circles. Some types use a gear mechanism and water pressure to generate a single stream or multiple streams. Stream rotors have relatively low precipitation rates, and multiple stream rotors provide matched precipitation for varying arc patterns.

submetering Separate metering of a portion of water use associated with a metered water-service connection.

subsurface drip irrigation The application of water via buried pipe and emitters, with flow rates measured in gallons per hour.

supplemental irrigation Application of water to a landscape in addition to natural precipitation.

system efficiency *See* hardware efficiency.

tall fescue A hybridized cool-season turfgrass characterized by deeper root systems and more drought tolerance than bluegrass.

thatch The buildup of organic material at the base of turfgrass leaf blades. Thatch repels water and reduces infiltration capacity.

turfgrass Hybridized grasses that, when regularly mowed, form a dense growth of leaf blades and roots.

undeveloped landscape *See* unlandscaped natural area.

unlandscaped natural area The portion of a development site where existing plant communities have not been removed or replaced.

valving Designation of irrigation circuits and corresponding remote-control valves.

vertical mowing Mechanical cultivation of turfgrass using vertically oriented knives to penetrate cored material, thatch, or underlying soil.

warm-season turfgrass Turfgrasses that grow vigorously in warm summer months, generally lose their green color, and are dormant in winter if the average air temperature drops below 50° to 60° F; some may die if exposed to subfreezing temperatures for extended periods.

wastewater Effluent.

water allowance *See* water budget.

water audit The on-site survey and measurement of hardware and management efficiency and the generation of recommendations to improve efficiency.

water banking A process whereby unused water allocations are added to future water allocations.

water budget The quantity of water needed to maintain plants and other features in an ornamental landscape.

water budget approach A method of establishing water-efficiency standards by describing limits on water consumption for irrigated landscapes.

water feature A fountain, pool, water sculpture, canal, channel, waterfall, constructed ponds and lakes, or other elements using water as part of a design composition.

water harvesting Design for capturing and using water runoff from storms on-site.

water-efficient landscape An ornamental landscape that minimizes water requirements and consumption through proper design, installation, and management.

water-service regulation Water-service contingencies usually established by water purveyors.

Index

NOTE: An *f* following a page number refers to a figure; a *t* refers to a table.

AD. *See* Allowable depletion
Adjusted water budgets, 51–52, 97, 163
Adjustment factors (AF), 50–51, 52, 79, 163
Administrative capacity, 18, 20, 78–79
Aerial photos, 44, 74
Aerification. *See* Coring
Aesthetic values, 2, 8
    demands of, 13, 16, 53
AF. *See* Adjustment factors
Agricultural crops, ET_c rates for, 44
Agriculture, water efficiency and, 21
Allocation, pooled, 98
Allowable depletion (AD), 100
Allowance. *See* Water budgets
Annual gallons, calculating, 79
APF. *See* Average plant factor
Applicability, increased range of, 18, 19, 20, 21
Application rate, 54, 71, 78
    description of, 64–65, 163
    metric conversion of, 101
Area, 163
    determining, 41, 43–44
Area takeoffs, 43–44, 79, 163
Arid climates, 163
    designing for, 2
    historic ET_O in, 45
    irrigation in, 4, 82
    water budget approach and, 40
As-built plans, 74, 163
Audits. *See* Water audits
Automated irrigation, 61–62, 163
    mismanagement of, 4, 16
    maintenance and, 69
Average plant factor (APF), 97
    determining, 50, 51, 52
    plant selection and, 50
AWWA Xeriscape Committee, guidelines from, v

Banking. *See* Water banking
Beneficial rainfall. *See* Effective precipitation
Blaney–Criddle formula, 82
Bluegrass, 163
    drought tolerance of, 58
    sprinkler irrigating, 48*f*
    See also Tall fescue; Turfgrass
Botanical gardens, irrigation of, 20
Bubblers, 63, 163
Budgets. *See* Water budgets
Building footprint, 163

California Irrigation Management Information System (CIMIS), information from, 81
Catch-can test, 46, 163

Ccf, 163
    conversion to, 49
Central irrigation control, 61–62, 163
Checklist approach, 7, 9, 50
    advantages of, 10*t*, 12, 15
    description of, 8, 10*t*, 12–13, 163
    design flexibility and, 15*t*
    disadvantages of, 10*t*, 13, 15*t*
    educational component of, 15*t*
    enforcement of, 10*t*, 15*t*, 78–80
    examples of, 103, 104
    implementing, 10*t*, 12, 18, 53, 73, 76–77
    management and, 15*t*
    metering and, 58
    opportunities for, 15*t*
    water budget approach and, 12, 80, 103, 125
CIMIS. *See* California Irrigation Management Information System
Clay, water-holding capacity of, 99
Clay loam, water-holding capacity of, 99
Climate
    variations in, 6
    See also Arid climates; Mediterranean climates; Semiarid climates
Clippings
    retaining, 68
    See also Mulch
Commercial landscape, 163
Compaction, alleviating, 68, 69
Compliance
    calculating, 69, 74, 78, 79, 79*f*, 80
    facilitating, 12, 62, 67, 73, 76, 77, 80
    improvements in, 14
    inspecting for, 19, 77
Compost. *See* Mulch
Connection fees, 22, 58
Conservation, v
    benefits of, 3, 17
    efficiency and, 4
    incentives for, 4–5, 14, 75
    local opportunities for, 4, 15, 15*t*, 19
    long-term, 12
    maintenance and, 67
    plant factors and, 51
    range of applicability and, 19
    scheduling and, 70
    short-term, 3, 9
    turfgrass and, 58
Consultants, working with, 13, 78, 80
Consumption
    limits on, 4, 7, 8, 9, 12, 19, 52
    monitoring, 8, 15, 53
    pooled, 98
Controllers, 165
    multiple-program, 61

multiple-start-time, 61
overriding, 59
programming, 70–71
Conversion factors, 41, 95–96, 163
Coring, 68–69, 164
Cost-effectiveness, 7, 20, 73
Coverage, head-to-head, 64f
Crop coefficient (K_c), 44, 164
Cumulative depletion, 97, 98
Cutting height. *See* Mowing height

Daily cycles. *See* Irrigation cycles
Demand, increase in, v, 1, 2–3, 73
Density, 45, 52, 71
Design, 64–65, 66–67, 71
   appropriate, 4–5
   checklist approach and, 53
   criteria for, 7, 8, 12, 52
   efficient, 3, 4, 7, 12, 43
   inefficient, 2, 54
   limits on, 2, 16
   monetary benefits of, 7
   standards for, 53–58
   understanding, 2, 14
   See also Functional design
Design concept statement, 77
Design flexibility
   checklist approach and, 15
   educational approach and, 15
   reducing, 13, 53, 56–57
   water budget approach and, 9, 15
Design review, using, 78
Detention basins, creating, 56
Development, small-scale, 13, 16
Distribution uniformity (DU), 164
   measuring, 46
   regulation of, 65
Drainage
   improving, 54, 55
   low head, 166
   storm, 55, 168
Drip irrigation, 164, 168
   advances in, 6
   damage and vandalism to, 66–67
   design criteria for, 66–67
   efficiency of, 47, 49f, 64
   hardware efficiency of, 46, 47
   runoff and, 64
   turfgrass and, 63–64
   using, 13, 51, 56, 62, 63, 64
   See also Sprinkler irrigation
Drought, 164
   conservation and, 9
   impact of, 4, 40, 41, 45
   supplemental irrigation and, 41
   See also Shortages
Drought tolerance, 13, 58
   reducing, 70
DU. *See* Distribution uniformity
Dual/multiple programming, 61, 164
Duration, 71, 78
   calculating, 61, 70, 101
   See also Frequency

Economic values, 8, 16
Edible crops, water efficiency and, 21
Educational approach, 7, 8, 13–14
   advantages of, 11t, 14
   design flexibility and, 15
   disadvantages of, 11t, 14
   effectiveness of, 14
   enforcement of, 11t, 15
   implementing, 11t, 15
   management and, 15
   voluntary participation in, 14
Educational programs, 2, 17, 56, 57, 73, 76
   checklist approach and, 12
   implementing, 3, 7
   importance of, 15, 16, 80–81
   incentives for, 4, 19
   problems with, 14
   regulation and, 19
   water audits and, 80
Education gap, addressing, 2, 6, 80
Effective precipitation, 45, 46, 97–98, 164
   impact of, 39, 52, 163
   See also Precipitation; Rainfall
Efficiency, v, 21, 62
   checking, 7, 12, 46, 78
   checklist approach and, 12
   conservation and, 4
   hardware, 46, 47, 165
   increasing, 1–4, 6, 7, 12, 14, 16, 20, 75, 80
   intrinsic benefits of, 14, 17
   local conditions and, 7
   long-term, 41
   management, 46, 47, 80
   standards for, 2, 19, 164
   See also Irrigation efficiency; Water-efficient landscapes
Emitters, 164
   drip, 62
   multiple-outlet, 67, 67f
   point-source, 167
   pressure-compensating, 66, 167
Enforcement mechanisms, 9, 76
Environmental benefits, v, 8
Erosion, reducing, 58
ET. *See* Evapotranspiration
ET_c rates. *See* Evapotranspiration
ET_O. *See* Reference evapotranspiration
Evaporation, 39, 71
   losses from, 64
   reducing, 68, 69
Evapotranspiration (ET), 4, 9, 71, 79, 164
   calculating, 44, 48–49, 51, 96
   estimated daily, 100
   information about, 81
   irrigation scheduling and, 39
   metric conversion of, 101
   precipitation and, 40, 40f, 82, 98
   real-time, 62
   scheduling and, 62, 70
   shading and, 70
Existing sites, 21
   irrigating, 43

Factored pooling, 75–76, 98, 164
Fertilizers, over-irrigation and, 69
Filters, 66, 69
Flooding, 2, 19
Flowmeters, 62
Flow rate, 62, 63, 66, 164
    metric conversion for, 102
Flow restrictors, 76
Frequency, 71, 78, 100
    setting, 61, 70
    See also Duration
Functional design, 2
    aesthetic benefits and, 13
    examples of, 53–54
    See also Design
Functionally required areas, 164
    designing, 53

Gallons per hour (gph), 164
Gallons per minute (gpm), 164
    measuring in, 63, 64
Government agencies. *See* Public authorities
gph. *See* Gallons per hour
gpm. *See* Gallons per minute
Grading, 53, 78, 164, 165
    complications from, 13
    design and, 7, 12
    plant selection and, 56
    regulation of, 55
    site, 55–56
    slope and, 44
Grass
    drought-tolerant, 13
    See also Bluegrass; Clippings; Tall fescue; Turfgrass
Green industry, 165
Ground covers, drip irrigation for, 16
Groundwater, 21
Grouping, 45, 52, 53, 71, 79
    importance of, 7, 56–57
    problems with, 64
    by water requirements, 57
    valving for, 57
Guidelines, v, 2
    applying, 3–4, 7, 18
    developing, 6, 8, 10t–11t
    See also Regulations

Hand watering, 59
Hardscape, 165
    IR and, 56
Hardware efficiency, 41, 71, 165
    calculating, 46
    drip irrigation and, 46, 47
    improvements in, 9, 61, 80
    IE and, 46, 47
    See also Efficiency
Head radius, 165
Head-to-head spacing, 64, 165
    matched precipitation rates and, 65
High-water-use plants, 78, 165
    APF for, 51
    ET_O for, 44
    irrigating, 75
    limiting, 53, 54, 56
    in nonirrigated areas, 50
    See also Low-water-use plants
Historic data, 41, 165
Horticultural practices, 7, 78
    water efficiency and, 69
Hydrozones, 57, 60f, 165
    analysis of, 78, 79
    scheduling and, 59, 61
    See also Irrigated areas

IA. *See* Irrigated areas
IE. *See* Irrigation efficiency
IF. *See* Frequency
Impact heads, 63, 65, 165
Inches per hour (in./h), measuring in, 64
Industrial landscapes, 165
    water efficiency and, 20
Inefficiencies, dealing with, 4, 15
Infiltration rate (IR), 71, 78, 165
    concern for, 54
    factors of, 55–56
    increasing, 56, 58, 68
    metric conversion of, 101
    precipitation rate and, 65
    reducing, 69
    soil texture and, 55t, 55–56, 61
    soil type and, 55t, 55–56, 61
Informational materials and services. *See* Educational programs
in./h *See* Inches per hour
Inspections, 73, 77–78
    compliance and, 76, 77
    by landscape professionals, 13
    on-site, 13
IR. *See* Infiltration rate
Irregular forms, irrigating for, 54
Irrigated areas (IA), 9, 43f, 165
    determining, 41, 54, 74, 79
    as variable factor, 52
    See also Hydrozones
Irrigation, 1, 39, 78
    in arid climates, 4, 82
    automated, 4, 16, 61–62, 69, 163
    concentrating, 43
    design and, 7
    efficiency in, 6, 21
    inefficiency in, 4, 6, 41
    metering, 9, 19
    mixed plantings and, 64
    relying on, 4
    See also Drip irrigation; Irrigation efficiency; Over-irrigation; Spray irrigation; Sprinkler irrigation; Supplemental irrigation
Irrigation circuits, 60f, 165
Irrigation controllers. *See* Controllers
Irrigation cycles, 71, 78, 165
    calculating, 65, 101
    multiple, 61
Irrigation efficiency (IE), 9, 41, 51, 79, 165
    accounting for, 49, 52
    hardware efficiency and, 46, 47
    management efficiency and, 46

measuring, 47, 96
Irrigation formulas, calculating, 70–71
Irrigation frequency (IF). *See* Frequency
Irrigation plans. *See* Plans
Irrigation professionals
    review and inspection process and, 76
    standards development by, 16
Irrigation scheduling. *See* Scheduling
Irrigation systems, 63–64
    design and, 12
    maintenance of, 69
    upgrades for, 14

K_c. *See* Crop coefficient

LA. *See* Landscaped area
Landscape coefficient, 71, 96, 166
Landscaped area (LA), 42*f*, 166
    determining, 41, 74
    water budget and, 43
Landscape professionals
    educating, 3, 4, 6, 13, 14, 77
    liability for, 77
    management efficiency and, 47
    review and inspection process and, 13, 77, 80
    standards development by, 16
Landscapes
    commercial, 163
    industrial, 20, 165
    mature, 45, 70, 80
    public, 20, 167
    rehabilitated, 21
    residential, 19
    See also Ornamental landscapes
Landscaping
    developer-provided, 19, 164
    homeowner, 20
    inefficient, 6
    nonfunctional, 166
    water-efficient alternatives in, 4
Lawn area, metric conversion of, 101
Limits, 166
    determining, 41, 43
    local, 3
Loam, water-holding capacity of, 99
Looped pooling, 75, 98, 166
Low-volume spray heads, coverage by, 65
Low-water-use plants, 45, 166
    hybridization of, 13
    irrigation and, 70
    sustaining, 51
    using, 54, 56
    See also High-water-use plants
Low head drainage, 166

Maintenance, 12, 16, 17
    checking, 47, 78
    conservation and, 67
    grounds, 7
    long-term costs of, 2
    management and, 7, 47
    standards for, 67–70
    system, 69
Maintenance schedules, establishing, 67, 69

Management, v, 54, 71
    checklist approach and, 12, 13, 15*t*, 53
    educational approach and, 15*t*
    improving, 2, 3, 4, 9, 59, 61
    inefficient, 57
    long-term, 3
    standards for, 7, 8, 16
    waste and, 16
    water budget approach and, 15*t*
    workshops for, 81
Management efficiency, 6, 7, 9, 41, 166
    changes in, 80
    IE and, 46
    maintenance and, 47
    scheduling and, 47
    See also Efficiency
Matched precipitation rates, 65, 166
Matched precipitation rate sprinklers, 61, 66*f*
Mature landscapes, 45
    water for, 70, 80
Mediterranean climates, 45, 166
Medium-water-use plants, 166
    ET_O for, 44
Metering, 19, 53, 62, 80, 95
    benefits of, 58
    dedicated, 9, 12, 58–59, 164
    looped, 75
    resistance to, 58–59
    See also Monitoring; Submetering
Methodology
    development of, 6, 71
    facilitating, 4
    scheduling, 71
    water-efficient, 2
Metric conversions, list of, 101–102
Microclimates, 45, 52, 70, 71, 166
    complications from, 13
    water requirements in, 57
Mixed plantings, problems with, 64
Model homes, designed landscape for, 19–20
Moisture-holding capacity. *See* Water-holding capacity
Monitoring, 53
    dedicated, 12, 58–59
    water budget approach and, 9
    See also Metering
Mowing height
    metric conversion for, 102
    optimum, 68
Mulch, 166
    generating, 68, 69
    runoff and, 69
Multiple-program controllers, 61
Multiple programming, 164
    advantage of, 62
Multiple-start-time controllers, 61
Multiple start times, 166
    scheduling, 65

Nonfunctional landscaping, 166
Nonirrigated areas
    high-water-use plants in, 50
    preservation of, 43, 51

Operating pressure, 166
  regulations on, 62
Ornamental landscapes, 53
  efficiency requirements and, 21
  ET and, 39
  irrigation scheduling for, 71
  water requirements for, 40, 57
  See also Landscapes
Overhead irrigation. *See* Sprinkler irrigation
Over-irrigation, 20, 57
  fertilizers and, 69
  problems with, 2, 6
  See also Irrigation
Overspray, 19, 46, 62, 166
  problems with, 2, 54, 64

Parks, irrigation of, 20
Percent switches, 166
  reprogramming, 61
Permits, 14, 73, 76
  landscape, 21
Pest control
  reducing, 17
  water requirements and, 70
pH, concern for, 54
Planimeter, using, 43, 74
Plans, 44, 165
  analysis of, 70, 77–78
  as-built, 74, 163
  planting, 12, 78, 167
  two-dimensional, 44
Plant factors (K_c), 71, 167
  conservation goals and, 51
  deriving, 79
  weighted, 96
Plant grouping. *See* Grouping
Planting plans, 78, 167
  design and, 12
Plants
  adapted, 166
  hybridization of, 13
  local, 17, 54, 57, 166
  salt-tolerant, 21
  See also High-water-use plants; Low-water-use plants; Medium-water-use plants
Plant selection, 15, 53, 55
  APF and, 50, 50t
  design and, 7
  grading and, 56
  problems with, 6
  regulating, 56–57
Plan view, 44, 167
Pooled allocation, 98
Pooled consumption, 98
Pooling, 167
  factored, 75–76, 98, 164
  looped, 75, 98, 166
  water budget approach and, 75
PR. *See* Precipitation rate
Precipitation
  average annual, 5f
  effective, 97–98, 164
  ET and, 40, 98
  variations in, 45
  See also Effective precipitation; Rainfall
Precipitation rate (PR), 64–65, 78, 167
  general formula for, 99
  IR and, 65
  matched, 65, 166
Pressure, operating, 62, 166
Pressure loss, 167
Pressure meters, 62
Pressure reducers, 62, 167
Professionals. *See* Irrigation professionals; Landscape professionals
Programming, multiple, 61, 62, 65, 164, 166
Pruning, 70
Public authorities
  conservation by, 4
  educating, 3
  water-efficient landscaping and, 20
Public awareness. *See* Educational programs
Public landscapes, 167
  irrigation of, 20

Rainfall (RF), 41, 82
  corrections for, 45
  effective, 45, 46
  ET and, 40, 82
  impact of, 39
  map of, 5
  supplemental irrigation and, 45
  See also Effective precipitation; Precipitation
Rain shutoff devices, 13, 59, 167
Range of applicability, 18, 19, 20, 21
Rate structures, changing, 22, 76
RCV. *See* Valves, remote-control
Reclaimed water, 167
  potential for, 21–22
Recreational areas, irrigation of, 20
Reference evapotranspiration (ET_O), 44, 50, 96, 167
  adjustment factor, 97
  historical data, 45
  rates, 47, 51
  See also Evapotranspiration
Regular forms, irrigating for, 54
Regulations, 62
  accepting, 3, 4, 7, 19
  alternatives to, 8
  developing, 6
  educational programs and, 14, 19
  local variation in, 6
  responsibility for, 73
  unenforceable, 20
  water-service, 76, 169
  See also Guidelines
Regulators. *See* Public authorities
Regulatory programs, 15, 18, 73
  enforcing, 76
Rehabilitated landscape, 21
Renovations, permits for, 21
Requirements. *See* Water requirements
Residential landscapes, irrigation of, 19

Resources, increased demand for, 2–3
Retrofitting, 46
RF. *See* Rainfall
Rigid polyvinyl chloride pipe, 67, 67*f*
Root systems, 45, 71
    improving, 58
Root zones, wetting, 46, 47, 66
Runoff, 39, 46, 167
    drip irrigation and, 64
    problems with, 2, 4, 56, 64, 68
    reducing, 55, 56, 61, 65, 69
    scheduling and, 56
    slope and, 56

Sand, water-holding capacity of, 99
Sandy loam, water-holding capacity of, 99
Savings, 1
    limits on, 20
    maximizing, 4, 18, 21
Scheduling, 16, 17, 48–49, 54, 67, 80
    calculations for, 45, 46, 82, 95–102
    checking, 12, 78
    conservation and, 70
    elimination of, 59
    ET data and, 39, 62, 70
    hydrozoning and, 59
    improvements in, 47, 62, 70–71
    long-term, 6, 78
    management and, 7, 47
    problems with, 65, 71
    runoff and, 56
    workshops on, 81
Semiarid climates, 167
    designing for, 2
    historic ET_O in, 45
Shading, ET rates and, 70
Shortages
    limits during, 4, 41
    public awareness and, v, 3
    See also Drought
Silt loam, water-holding capacity of, 99
Simple water budgets, 167
    formula for, 97
Site areas, 42*f*, 167
    measuring, 41, 44
Site inspection, 67, 70, 80
Site surveys, 74
    determining slope with, 44
Slope, 71
    determining, 44
    minimizing, 55
    runoff and, 56
Small projects
    checklist approach and, 13
    standards and, 16
Software, irrigation scheduling, 71, 80
Soil improvement, 53, 54–55, 78, 167
    design and, 12
Soil-moisture sensors, 59, 66
    maintaining, 61, 69
Soil reports, 55, 78, 167
Soil texture, IR and, 55
Soil types, 7, 71
    concern for, 54

IR and, 55
variation in, 13
Spacing, head-to-head, 165
Spray heads, 63, 167
    low-volume, 65
Spray irrigation, 168
    alternatives to, 56
    efficiency of, 16, 46, 47
    operating pressure and, 62
    See also Irrigation
Spray patterns, regular, 54
Spreadsheets, working with, 75, 80
Sprinkler heads, 54, 69
Sprinkler irrigation, 54, 59, 63, 168
    restricting, 56
    design criteria for, 64–65
    efficiency of, 6, 46, 47, 48*f*, 64
    inefficiency in, 4
    turfgrass and, 57
    See also Drip irrigation; Irrigation
Sprinkler patterns
    adjusting, 79
    PR for, 99
Sprinklers, matched precipitation rate, 61, 66*f*
Standardization, 13
Standards, 6, 103–104
    AF and, 51
    applying, 16, 18, 20–21
    design, 53–58
    developing, v, 2–4, 8–9, 15–17, 46–47, 73
    educational component of, 17
    enforcement of, 18
    irrigation, 58–67
    maintenance, 67–70
    management, 47
    scheduling, 70–71
    small projects and, 13, 16
    state resources for, 159–160
Storm drainage, 55, 168
Storm water, on-site management of, 56
Strainers, cleaning, 69
Stream rotors, 63, 65, 168
Submetering, 9, 168
    complications from, 12
    dedicated, 58–59
    See also Metering
Submittal requirements, 77–78
Subsidies, 22
Subsoil, IR and, 56
Subsurface drip irrigation. *See* Drip irrigation
Subsurface water quality, 46
    improving, 58
Supplemental irrigation, 46, 54, 82, 168
    drought and, 41
    increasing, 75
    need for, 40, 97, 98
    rainfall and, 45
    See also Irrigation
Surface area, metric conversion for, 101
Surface compaction, alleviating, 68, 69
Surveys. *See* Site surveys
System-capacity charges, 58

System efficiency. *See* Hardware efficiency
System upgrade services, promotion of, 80

Tall fescue, 168
   drought tolerance of, 58
   See also Bluegrass; Turfgrass
Technology
   facilitating, 4, 6
   impediments to, 13
Thatch, 168
   removing, 68, 69
Thinning, 70
Top soil, preserving, 55, 56
Toro Company, data from, 82–94
Total planted area, APF for, 51
Transpiration, 39, 70
Turfgrass, 56, 167
   benefits of, 57–58
   cool-season, 44, 164
   DU for, 65
   drip irrigation and, 16, 63–64
   ET rates on, 44, 46
   fertilizer for, 69
   head-to-head spacing and, 64
   hybridized, 58
   irrigation for, 6, 20, 45*f*, 48*f*
   limitations on, 13, 16, 53, 57–58, 78, 166
   maintenance of, 9, 47, 50, 68–69
   mowing height for, 68, 68*t*
   warm-season, 168
   See also Bluegrass; Tall fescue
Turf perimeter/area ratio, using, 58
Two-dimensional plans, 44

Undeveloped landscape. *See* Unlandscaped natural area
Uniformity, 46, 54
Unlandscaped natural area, 168

Valves, 74
   check, 62, 63*f*, 163
   flush, 62, 164
   automating, 59, 61–62
   installing, 62
   remote-control, 59, 61, 167
Valving, 57, 168
Vegetable gardens, water efficiency for, 37
Vertical mowing, 68–69, 168

Wastewater. *See* Reclaimed water
Water allowance. *See* Water budgets
Water audits, 9, 12, 16, 168
   checklist approach and, 13
   DU and, 46
   as educational tool, 80
   performing, 65, 67, 69, 78, 80
   promotion of, 81
   training for, 14, 17
Water banking, 75–76, 168
Water budget approach
   advantages of, 9, 10*t*, 15
   checklist approach and, 12, 103, 125
   criticism of, 46
   description of, 7, 8–9, 10*t*, 168
   design flexibility and, 15*t*
   disadvantages of, 9, 10*t*, 12
   educational component of, 15*t*
   enforcement of, 10*t*, 15*t*
   examples of, 103, 152
   implementing, 10*t*, 40, 73, 74–76
   management and, 15*t*
   opportunities for, 15*t*
   pooling and, 75
Water budgets
   adjusted, 51–52, 97, 163
   billing systems and, 75
   in Ccf, 97
   checklist approach and, 80
   empirical basis for, 46
   establishing, 9, 18, 43, 49, 51, 52, 74–75, 78, 80, 82, 95–102, 168
   formulas for, 39, 41, 49–52, 97
   metering for, 58
   negative image of, 12
   noncompliance with, 76
   simple, 97, 167
   site-specific, 8–9
   submitting, 51, 74–75
Water-conserving plants. *See* Low-water-use plants
Water delivery, metric conversion for, 102
Water efficiency. *See* Efficiency
Water-efficiency standards. *See* Standards
Water-efficient landscapes, 53, 169
   benefits of, 14
   designing, 54
   managing, 14
   resistance to, 19–20
   See also Efficiency
Water features, 169
   limiting, 53, 54, 56
Water harvesting, 56, 169
Water-holding capacity, 47, 56, 71
   improving, 54
   for various soil types, 99
Watering basins, constructing, 64
Watering window, 71
Water measurements, list of, 95
Water pooling. *See* Pooling
Water requirements, 40, 44, 166, 167
   actual, 41
   calculating, 8, 57, 95–102
   historic, 41
   mowing height and, 68
   pest control and, 70
   reducing, 3–4, 53
   seasonal variations in, 8, 51
Water shortages. *See* Shortages
Water use
   local patterns of, 15
   prioritizing, 75
Weed control, 17, 70
Wind drift, 62
Workshops, 81

Xeriscape Committee (AWWA), guidelines from, v